An

Introduction

to

SCR POWER CONTROLS

George A. Sites

BACKGROUND

THIS BOOK WAS WRITTEN TO HELP ENGINEERS, TECHNICIANS AND SALES
ENGINEERS WHO HAVE LITTLE OR NO SPECIAL TRAINING IN THE DESIGN
AND APPLICATIONS OF SCR POWER CONTROLS. THE INTENT IS TO AID IN
UNDERSTANDING THE BASICS AND NOT BE A "TEXT BOOK" IN THE TRUE
SENSE OF THE WORD. WITH MINIMAL REFERENCE TO TECHNICAL CONCEPTS
REQUIRED, IT SHOULD SERVE AS THE PERFECT PRIMER.

George A. Sites

FROM THE AUTHOR

I HAVE SPENT NEARLY MY ENTIRE PROFESSIONAL CAREER, OVER 30 YEARS SPECIALIZING IN THE DESIGN AND APPLICATIONS OF SCR POWER CONTROLS. MANY TIMES I HAVE BEEN ASKED TO RECOMMEND A BOOK ABOUT SCR POWER CONTROLS. HOWEVER, SINCE NEARLY EVERY BOOK AVAILABLE ON THE SUBJECT IS ABOUT THE DISCRETE SCR DEVICE AND NOT THE COMPLETE CONTROLLER, I DECIDED TO WRITE AN EASY TO UNDERSTAND BOOK.

HERE IT IS. I HOPE YOU FIND IT USEFUL!

DEDICATION and THANKS

I dedicate this book to the many engineers, technicians, assemblers, managers, sales engineers, and, of course, clients with whom I have worked over the years. Although the list is too long to detail in its entirety, I must give special thanks to Jeff Robinson and Chuck Denney with whom I started the SCR Power Control product line at HDR Power Systems and Bern Smith, our first engineer. Without their valuable help it could not have happened. Also a thank you to all the loyal and dedicated employees of HDR Power Systems who also helped make it all happen.

IN MEMORY OF

This book is written in memory of Marshall Clanin and Chuck Donnal. Both were former colleagues of mine at Halmar Electronics and close personal friends. Marshall Clanin was Vice President of Engineering and Chuck was Sales Manager. Each, in their own way, taught me very valuable lessons that I still use today. Although Marshall and Chuck have passed away, they are forever etched in my memory. I miss them both!

NOTES

Unless otherwise indicated, all drawings are by
Charles E. Denney, Jr.

Edited by Alida L. Breen

Special Assistance by Barbara Emery

Cover Designed by W.L. Cox

Copyright 2004

G.A. Sites Books
5653 Haydens Reserve Way
Hilliard, Ohio 43026 USA

While the author and the publishers believe that the information and guidance given in this book is correct, all parties must rely upon their own skill and judgment when making use of it. Neither the author nor the publishers assume any liability to anyone for any loss or damage caused by any error or omission in this book, whether such error or omission is the result of negligence or any other cause. Any and all such liability is disclaimed.

Table of Contents

LIST OF PHOTOGRAPHS AND ILLUSTRATIONS

Chapter 1

INTRODUCTION to SCR
POWER CONTROLS

INTRODUCTION

SCR Power Controls were first developed not long after the SCR -
in the late 1950's. Originally available with ratings of only a
few hundred amps, power controls have increased in current
ratings to thousands of amps. As the current ratings increased,
so have the applications. Now, with applications in nearly every
industrial process, it is clearly the product of choice for
controlling large amounts of power. Basically, any process
using electric heat is a possible application for SCR Power
Controls.

Each of the following chapters contains information specific to different areas of interest concerning SCR Power Controls. These include: applications, advantages/disadvantages, different firing methods, and load considerations, to name a few. If you want to learn more, read on.

First and foremost, SCR Power Controls provide an economic and reliable means for controlling large amounts of electrical power. Since it is solid-state and has no moving parts, the maintenance problems are nearly non existent. Also since no large magnetic devices such as a saturable core reactor is required, it is more efficient.

In addition, due to the high number of control options available, the SCR Power Control can be used in nearly every electric heating application.

Figure 1 – Two Leg Control SCR Power Control
(photo courtesy of HDR Power Systems, Inc.)

MAJOR ADVANTAGES

Low maintenance – with no moving parts to wear out, routine maintenance other than cleaning is not required.

High reliability - Because it is solid-state and there are no mechanical parts to wear out, years of reliable service should be expected.

Control options - With an SCR Power Control, depending upon the application, any of three different types of control can be specified: 1) zero-fired; 2) phase-fired or 3) zero-fired into a transformer. In addition, each of these has its own set of control options making the SCR Power Control extremely versatile.

Fast response - Compared to a variable transformer or many other electrical control devices, it can make large changes in output in very short time periods. This fast response allows for applications using loads with a wide varying resistance, it eliminates fluctuating line voltages as a problem, and it can respond to fast changing command signals.

Infinite resolution - Unlike the step changes in a transformer and tap switch combination, the resolution is nearly infinite, accurate and very fast.

Small size - In comparison to transformer/tap switch combinations or saturable core reactors, it is smaller.

DISADVANTAGES

Although the disadvantages are few compared to the advantages, there are four that need addressed 1) Power Factor, 2) Harmonics and 3) Radio Frequency Interference (commonly referred to as RFI) for Phase-fired equipment and 4) light flickering for Zero-firing. Light flickering is not very common and is discussed in the chapter on Zero-firing. Power Factor, Harmonics and RFI problems are more common and require a complete chapter.

3

SUMMARY

Most other types of electrical control devices do not offer the flexibility available with SCR Power Controls. This contributes to their popularity. The advantages are great in comparison to the disadvantages.

NOTES CHAPTER 1

Chapter 2

INDUSTRIES and APPLICATIONS

INTRODUCTION

Have you ever thought about how many industrial processes use heat as part of the process? It's a big number! Now think about how many don't use heat as part of the process. Actually, there are not very many.

Now imagine the cost of fossil fuels continuing to increase making electricity more attractive as an energy source to these industries. Nearly every industrial process using electric heat is a possible application for an SCR Power Control.

For me it's easy to draw a parallel between the SCR Power Control and the personal computer. The applications keep growing and growing while the equipment continues to become more powerful.

INDUSTRIES

There are six major industries that use the lion's share of SCR Power Controls. They are as follows: Composites, Converting, Metals, Pulp & Paper, Plastics and Rubber.

> **Composites** - Composed of the ceramic, fiber and glass industries because they are all very similar. The main process equipment used is autoclaves, dryers, extruders, furnaces, heater, kilns and ovens.

Figure 2 – Electric "Glory Hole" Furnace
(photo courtesy of Electroglass)

7

Converting - It is composed of the paper, paperboard, foil and film industries. They use coaters, curing equipment, dryers and metallizers.

Metals - Uses dryers, fluidized bed furnaces, ovens and RF induction. This is one of the smaller application areas.

Pulp & Paper - Applications include calendars, sheet drying and pulp molding.

Plastics - These generally are low current applications but high in quantity. Applications include autoclaves, calendars, dryers, extruders, ovens, pre-heaters and sealing/welding equipment.

Rubber - Similar to the plastics industry, applications include autoclave, calendars, curing equipment, dryers extruders, furnaces, molding ovens and microwave systems.

APPLICATIONS

The number of applications continues to grow and grow all the time. Here is a partial list of equipment types that utilize SCR Power Controls.

* Autoclaves
* Chemical Heating
* Dryers
* Electrode Manufacturing
* Environmental Chambers
* Extruder and Forming Equipment
* Furnaces
* Glass Manufacturing
* Heat Exchangers & Hot Oil Systems
* Induction Heating
* Kilns
* Lamp Heating
* Lighting (airports, theaters, etc.)
* Microwave Dryers & Heaters

* Ovens
* Resistive Heating
* Semiconductor Processing
* Vibratory Feeders
* Wire Annealing

Each of these categories can be expanded further. For example, some of the applications include:

Dryers - Air, Gas, Liquid - Atmospheric, Batch, Blood, Brick, Clay, Ceramic, Porcelain, Carpet & Rug, Centrifugal, Chemical, Coal, Continuous, Convection, Cylindrical, Fabric, Film, Fluid Bed, Food, Fragrance, Glove, High Temperature Air, Infrared, Ink, Laboratory, Microwave, Pharmaceutical, Sand & Gravel, Sludge & Slurry, Textile, Tobacco, Tunnel, Ultraviolet and Vacuum.

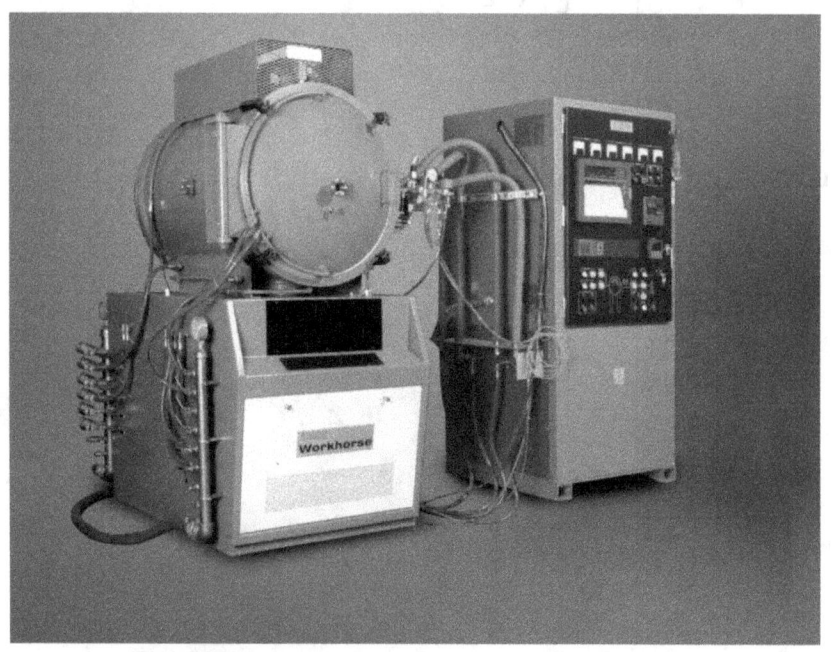

Figure 3 – Centorr Vacuum Industries Workhorse ™ II Brazing, Heat Treating and Annealing Furnace with High Vacuum Pumping System, Designed for Maximum use of 1315C
(photo courtesy of Centorr Vacuum Industries)

Furnaces - Accurate Temperature, Aluminizing, Aluminum Alloy Treatment, Aluminum Melting, Annealing, Hardening, Batch, Bench, Billet Heating, Blast, Box Type, Brass Melting, Bright Annealing, Calcining, Carbonitriding, Carbonizing, Carburizing, Ceramic, Chemical Vapor Deposition (CVD), Clay, Continuous Annealing, Copper Melting, Crystal Growing, Descaling, Diffusion, Electric Arc, Enameling, Extrusion, Glass, Glass Tempering, Glazing, Gold & Silver Melting, Salt Bath, Hardening & Drawing, Heat Treating, High Temperature, High Vacuum, Homogenizing, Hot Pressing, Induction Heating, Isothermal Heat Treating, Lead & Magnesium Melting, Nitriding, Normalizing, Sintering, Smelting, Soaking, Pit, Solar and Stress Relieving.

Figure 4 – Portable MeltMizer ™ is a complete yet compact high temperature furnace facility that enables scientific evaluation of a broad range of process conditions.
(photo courtesy of Diversified Controls & Systems, Inc.)

Glass - Annealing, bending, Float Lines, Forehearth, Lehr, Melting, Metallizing, Mirrors and Molding.

Kilns - Brick & Clay, Carbonizing, Cement, Chemical, Ceramic, China, Pottery, Tile, Continuous, Drying, Lime Burning, Lumber, Sintering, Sludge Burning, Super Conductors and Tunnel Kilns.

Lamp Heating - Quartz, Tungsten and Infrared.

Ovens - Aging, Annealing, Armature, Bakeout, Bakery, Batch, Bonding, Brick, Calcining, Cereal Roasting, Clean room, Coke, Conditioning, Core Baking & Mold Drying, Crystal Growing, Curing, Dehydrating, Dielectric, Enameling, Lacquering, Food
Processing, Galvanizing, Glass, Heat Treating, Humidity Controlled, Infrared, Microwave, Paint Drying, Pharmaceutical, Powder Coating, Preheating, Radiant Heat, Rubber Curing, Shrink Film, Sterilizing, Stress Relieving, Tempering, Textile, Tunnel, Ultraviolet and Vacuum.

SUMMARY

As you can see, the applications are endless. Virtually any industrial process that uses electric heat is a possible application for an SCR Power Control.

Chapter 3

MAIN COMPONENTS of an
SCR POWER CONTROLLER

GENERAL

An SCR Power Control consists of the following main components.

* Silicon Controlled Rectifiers (SCR)
* Heatsink
* Firing Circuit
* Protective Devices
* Optional Regulation Circuits (covered in another chapter)

SILICON CONTROLLED RECTIFIER (SCR)

The SCR is the solid-state power-handling device. An SCR conducts current in only one direction and has two states - either on or off. Because the SCR only conducts in one direction, two SCRs are connected in an inverse parallel configuration or more commonly referred to as "back to back" connected.

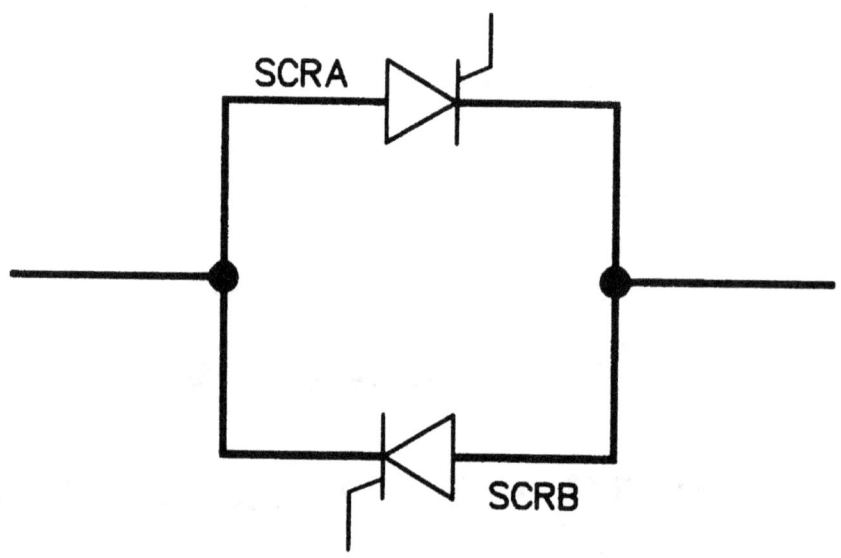

Figure 5 - Schematic of two SCRs connected "back to back".

By gating the two SCRs ON alternately, both half cycles of AC power is switched. This assembly is most commonly referred to as an AC Switch.

All modern SCR Power Controls use two SCRs in each controlled leg of power; for example a single phase power control would require one AC Switch assembly and a three phase unit would require either 2 or 3 (See Chapter 5 for more details).

HEATSINK

The heatsink carries the heat away from the SCR allowing it to operate within its temperature ratings. Nearly all electrical ratings of an SCR are degraded as the operating temperature increases. Since SCRs dissipate approximately 1.5 watts per ampere of controlled current, it is easy to understand why excessive temperatures cause most failures to SCRs.

Since many times the heatsink is the largest component, it basically determines the overall size of the SCR Power Controller.

FIRING CIRCUIT

The firing circuit is an electronic circuit that controls the operation of the SCRs. The output power level is controlled in proportion to an input command signal. The command signal can be from any common process controller with an output such as 4-20 ma or something as simple as a potentiometer.

This circuit also provides the isolation between the main power and the command signal. This is accomplished by using isolation transformers for control power, optical isolators for command signal inputs and pulse transformers that galvanically isolate the control circuit from the main power. All of this isolation helps prevent unwanted transient signals from falsely firing the SCRs.

The firing circuit provides gate signals that turn on the SCRs. The timing of these pulses is critical. Since the SCR resets when the current through it is a zero, there is no need for the firing circuit to turn the SCRs off. In the case of a simple AC switch (two SCRs connected back to back) if the positive and negative half cycle SCRs are not turned on at precisely the correct time, then a DC component is generated. This DC component can cause power transformers to overheat due to core saturation. In the case of a three-phase unit, the timing of

Figure 6 – Three-Phase, Phase Fired Firing Circuit
(photo courtesy of HDR Power Systems, Inc.)

the gate pulses to each phase must be precisely 120 degrees apart to maintain balanced currents in the three-phase mains. Digital Phase-Lock-Loop timing circuits are the most accurate and should be used whenever possible.

PROTECTIVE DEVICES

There are basically two types of SCR protection utilized. Line fuses are used for over current protection and Snubber/Transient protection circuits help prevent unwanted transient voltage spikes from mis-firing the SCRs.

Semiconductor Fuse

The semiconductor fuse is very fast acting and is commonly referred to as an I^2T fuse. Besides the voltage and current ratings of the fuse, the I^2T rating plays an important role. For proper protection of the SCR coordination of the SCR's and fuse's I^2T ratings are critical. The I^2T of the fuse must be smaller than the SCR's in order to protect the SCR.

The semiconductor fuse operates at a high temperature (over 100°C) in order to make it fast acting. As with all fuses, steady state operation of the fuse should be at approximately 80% of its current rating.

Many of the modern semiconductor fuses have a square body instead of the round body that's been in use for many years. This is simply the result of evolution in fuse design. Manufactures of square body fuses claim to have the same I^2T ratings as round fuses but in much smaller packages. The cost of the square body fuse is also lower.

It is important to note that semiconductor fuses are intended to protect the SCRs, not the load. It is the user's responsibility to provide the proper fusing for the particular load in accordance with state and local codes.

Snubber/Transient Suppression Circuits

The snubber circuit is an R-C circuit (resistor in series with a capacitor) and is sized to "slow down" any transient to which the SCR power control is exposed. The R-C time constant is selected to keep transient voltages to a level less than the dv/dt rating of the SCR and prevent false turn on of an SCR. It is in parallel with the SCRs.

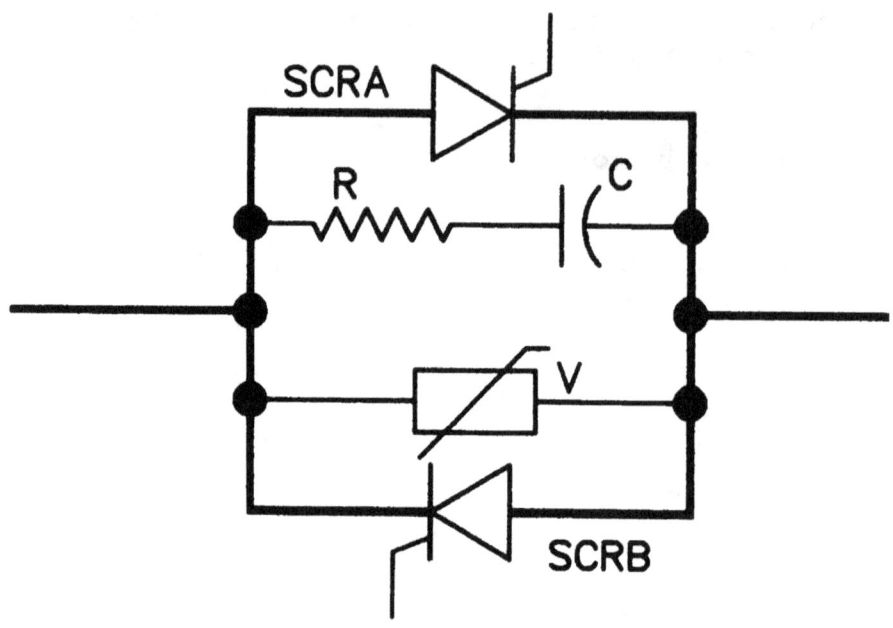

Figure 7 - Schematic of RC Snubber and MOV Transient Protection Circuit with Back-to-Back SCRs.

A Metal Oxide Varistor (MOV) is used to prevent unwanted transient voltages from exceeding the PIV rating of the SCR. The MOV has steady state, clipping voltage and power or joule ratings. Obviously the steady state rating must be higher than the expected maximum operating voltage. The clipping voltage must be below the PIV rating of the SCR to provide proper protection. The power or joule rating is picked based upon expected transient power levels or in some cases simply because the mechanical package makes more sense.

The MOV is available in many ratings, sizes and shapes. It is available as a lead mount, a bolt down package or even as a hockey puk. It is up to the designer to select the proper package type.

18

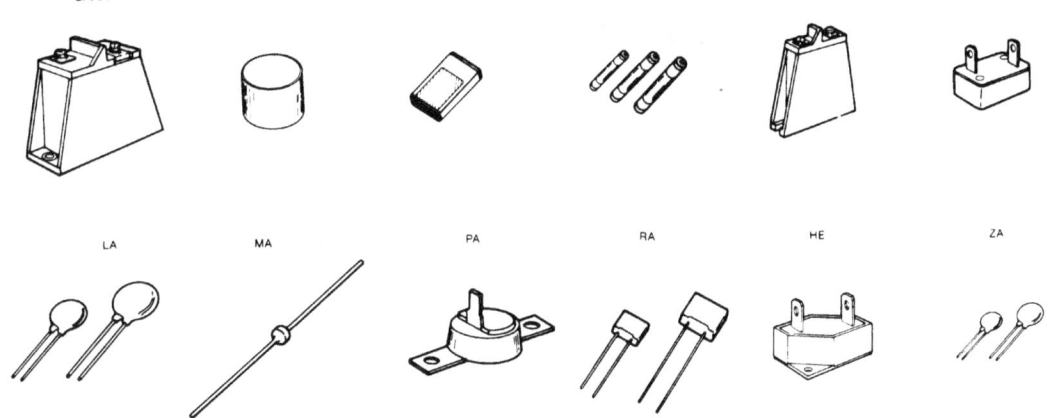

LA   MA   PA   RA   HE   ZA

Figure 8 - Samples of MOV Packages
(drawing courtesy of Harris Semiconductor)

SUMMARY

Although the SCR Power Controller is simple in function, it has
several components that can make or break the design. By
carefully selecting the components, a very reliable product can
be designed.

19

Chapter 4

SCRs - the DISCRETE DEVICE

INTRODUCTION TO THE DISCRETE SCR

The General Electric Company invented the Silicon Controlled Rectifier (SCR) in 1957. Although originally called the SCR, the name was not universally accepted and the name was changed to the Thyristor. However, in the United States, the term SCR is still most commonly used, while in other countries it is still referred to as a Thyristor. For the purposes of this book, I have used the term SCR.

Even though the SCR is over 40 years old, it is still the most efficient solidstate device for handling large power levels. Since it is available in a large variety of voltage/current ratings and mechanical configurations, it is one of the easiest solid-state power switching devices to use.

Today there are many manufacturers of SCRs. Some specialize in small sizes or faster switching speeds or even different mechanical configurations. Therefore, the engineer has many choices when designing.

The SCR is used in many electrical products including: SCR Power Controls, AC and DC Motor Drives, Alarm Clocks, Light Dimmers, Uninterruptible Power Supplies, Battery Chargers, Rectifiers, Garage Door Openers, etc. The list goes on and on. This book concentrates on the SCR Power Control.

THEORY OF OPERATION

The SCR is very much like a Diode (a device that conducts current in only one direction and cannot be controlled) except it has three terminals. The three terminals are the Anode (negative), the Cathode (positive) and the Gate (the control terminal). The SCR will only conduct current in one direction similar to the diode; however, the Gate "turns on" the SCR when it receives a firing pulse. As long as the pulse is present, the SCR remains on. When the pulse is removed, the SCR returns to the blocking state when the current through it goes to zero.

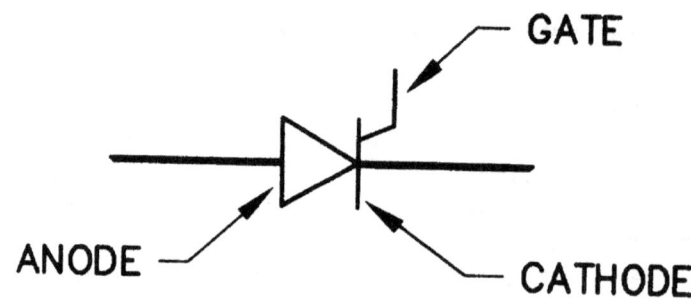

Figure 9 - Schematic Symbol for an SCR

FIRING THE SCR

Although there are many ways to turn on or fire an SCR, some are intentional and some are not. The most common and most desirable method of intentional firing the SCR is by putting a positive DC current on the gate terminal. By using this method you can precisely determine when the SCR will turn on.

There are two common types of gate pulses. The first type provides a single high-energy pulse or "hard fire" to the gate and the second provides a series of pulses or a "pulse train" to the gate.

Hard Firing

When hard firing an SCR, the firing circuit provides an initial high current pulse (usually several amps in magnitude) and follows up with a steady DC current of lower magnitude, usually referred to as the "back porch". This method initially turns the SCR on and relies upon the "back porch" to keep it turned on. The pulse occurs once each half cycle of AC current and works well until an unwanted, externally generated pulse turns the SCR off. When this happens, the SCR remains off until the next half cycle of current and the firing circuit provides another hard pulse. In some cases this is undesirable and can cause damage to the SCR or other externally connected equipment.

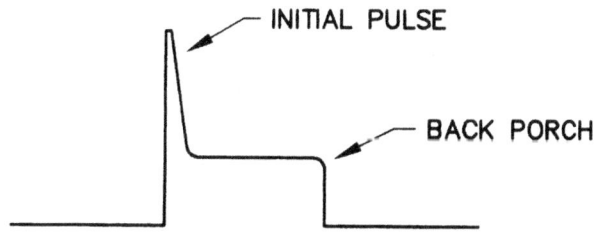

INITIAL PULSE

BACK PORCH

Figure 10 - Hard Firing Gate Pulse

23

Pulse Train Firing

Pulse train firing occurs when a long string of pulses is applied to the gate of an SCR. Usually each pulse is 1 or 2 amps in magnitude and is pulsed at a rate of 15 to 20 kilohertz. As you can see in Figure 11, the gate is pulsed numerous times. If the SCR is turned off by an unwanted external signal, the next firing pulse will turn the SCR on. There is no waiting until the next half cycle of current to turn the SCR back on.

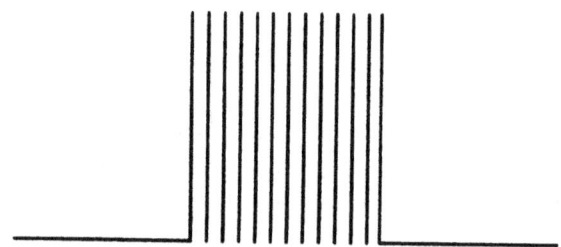

Figure 11 - Pulse Train Gate Pulse

Although I have used both methods, I have had the most success using the pulse train. However, with either method it is important that the unwanted, externally generated transient pulses be "snubbed" out. Circuits commonly referred to as Snubber Circuits help prevent the SCR from being falsely turned on or off. These circuits are covered in a later chapter.

One of the most significant improvements to the discrete SCR since its inception was the addition of the amplified Gate. The early SCRs required very large Gate currents to turn on which, in turn, required large power supplies as part of the firing circuit. The addition of the amplified Gate reduced the Gate current requirements and helped simplify the firing circuit design. To my knowledge, all SCRs produced today have amplified gates.

24

CHARACTERISTICS

Each of the following characteristics is important in selecting an SCR or in evaluating a problem. There are many more, but these are the most significant.

Gate Drive Requirements

Although a relatively small gate pulse will turn on an SCR (appx. 100 ma @ 3 v for 10 microseconds), it is very important to have plenty of gate drive. Turn on times and surge current ratings can be adversely affected if the gate drive is insufficient. Experience has shown that 24 Vdc @ 1 A will reliably turn on most SCRs.

di/dt

This is the rate of rise of anode current capability during turn on. Insufficient gate drive can cause SCR failures during turn on due to not turning the SCR fully on.

dv/dt

This is the rate of rise of forward voltage capability. Externally generated transients that exceed the dv/dt rating of the SCR can cause false firing of the SCR. Properly sized snubber circuits will help prevent this.

Peak Inverse Voltage (PIV)

This is the maximum peak voltage that can be applied to the SCR without the SCR turning on. This rating should be a minimum of 2.25 times the line voltage that will be applied to the SCR.

Holding Current or Latching Current

This is the minimum forward current that causes the SCR to maintain conduction. Although the holding and latching currents are slightly different (the latching current is usually higher), the terms are used interchangeably.

I^2T (Pronounced Eye Squared Tee)

This refers to the subcycle current rating of an SCR or semiconductor fuse. The coordination of this rating on both the SCR and the semiconductor fuse is important for the fuse to protect the SCR. The fuse rating needs to be lower than the SCRs.

Thermal Impedance

This is the rating for heat transfer from the SCR. It is expressed as degrees centigrade per watt of dissipation.

Current Rating

The rating is usually in RMS or Average current. It is the maximum current at which the SCR can be operated. Generally, the RMS rating is used in AC applications and the Average rating in DC applications.

Forward Drop

This is simply the voltage drop across the SCR during conduction. The voltage drop times the RMS current will give the power loss in watts of the SCR for that current level.

MECHANICAL PACKAGES

As mentioned earlier, the SCR is available in many different mechanical packages. Each is as different from the next as it is similar. Each of these could be used in an SCR Power Control. Mechanical lugs are not the types used on printed circuit boards.

There are four basic mechanical package types: Stud Mount; Hockey Puck; SCR Module and Solidstate Relay (SSR).

Stud Mount

This is the original power SCR package. Ratings of 600 to 1000 PIV and 300 Amps are common. Stud mount SCRs are not as commonly used today as they once were. Because they are single side cooled it is harder to remove heat and therefore the current ratings are limited. Packaging issues also limit the available PIV ratings.

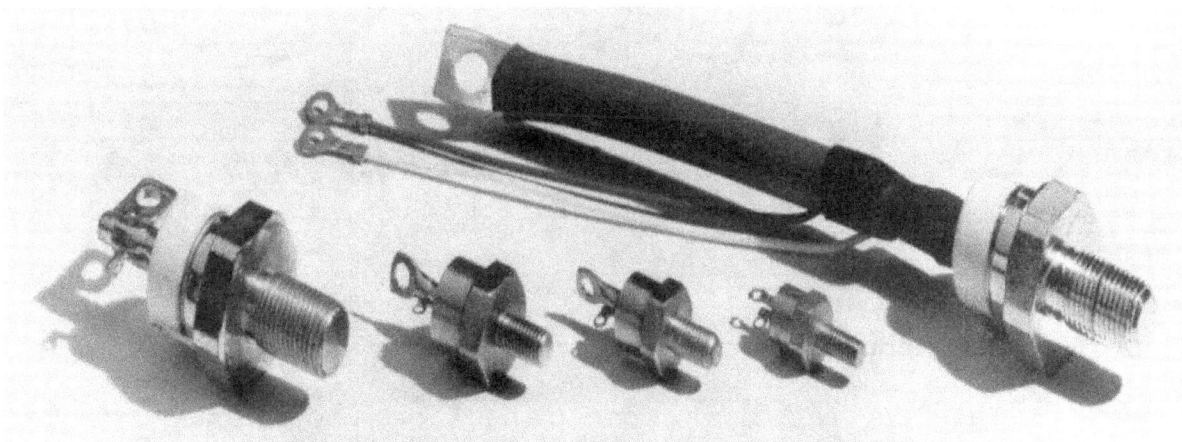

Figure 12 - Stud Mount SCRs
(photo courtesy of Richardson Electronics)

Hockey Puk

This is the most commonly used package for very high voltage and current ratings. The name (you guessed it) comes from the shape being similar to a hockey puk. Voltage and current ratings in the thousands of amps and /or voltage are common today. The high power handling capability of the hockey puk is due in part to being double side cooled when mounted between two heatsinks.

27

Figure 13 - Hockey Puk SCRs
(photo courtesy of Richardson Electronics)

SCR Module

This is the most common package used today for medium voltage and current ratings. Ratings of 2000 PIV and 350 A are common. What separates this from the stud mount and the hockey puk SCR is that the module contains two (2) SCRs. The Anode of one SCR is connected internally to the Cathode of the other SCR to simplify circuit configurations. This arrangement allows for the module to be used in an AC Switch or a DC Rectifier configuration. The current rating of the module refers to the average current rating of one SCR, not the entire package. The introduction of this package allowed designers to reduce the size and cost of medium current rated SCR Power Controls.

Figure 14 - Dual SCR Modules
(photo courtesy of Eupec)

Solid-state Relay (SSR)

This module is like the SCR Module since it has two SCRs but the similarity ends there. Unlike the SCR Module, the SSR has the two SCRs (connected in an inverse parallel configuration) and additional circuitry. This additional circuitry includes an RC Snubber and two optical isolators to drive the SCRs and optionally a zero-crossing detector. Two types of Solid-state Relays are available: zero-crossing and random turn-on (phase-firing). The difference

29

between the two types is the zero-crossing detector that is eliminated in the random turn-on device. Current ratings up to 100A and voltage ratings for operation at 575 Vac are commonly available. Another difference from the SCR module is the SSR's current rating is for the complete module and not the individual SCRs mounted inside.

Figure 15 - Solid-state Relays (SSR)
(photo courtesy of Crouzet)

HEATSINK SELECTION

As with any solid-state device it is extremely important to maintain as low an operating temperature as possible. Thermal Impedance of the SCR and Heatsink plus the watts loss curves of the SCR provide the information required to calculate temperature rise for any given steady state current rating.

Cooling can be passive by convection or as active as fans or cooling water would provide. As you change from passive to active cooling, larger current ratings can normally be achieved in much smaller package sizes.

Heatsinks come in many different sizes and shapes, usually designed with a certain SCR package in mind. When designing a new SCR and Heatsink combination, the engineer must first calculate the maximum allowable temperature rise for the SCR, the thermal impedance for the heatsink and, pick the heatsink based upon the SCR package type. Finally, you must determine whether it will have active cooling such as fans or water and the size or shape required due to mechanical restrictions. A custom shaped heatsink can be designed and manufactured at a reasonable cost if a standard designed heatsink does not meet the requirements.

Figure 16 - Common Air Cooled Heatsink
(photo courtesy of Crouzet)

MOUNTING THE SCR

Extreme care should be taken when mounting or making connections to an SCR. For SCR modules and Solid-state Relays, each of the mounting holes should be torqued according to the manufacture's specifications. The electrical connections tend to be fragile and care should be taken to ensure proper tightness without over tightening.

Hockey Puk SCRs require more care in mounting even though it is not quite as fragile. When installing an SCR between two heatsinks, a device called a Clamp is used to hold both heatsinks and the SCR together as an assembly. Clamps are selected based upon clamping pressure, SCR diameter and the length of both required. Since the Silicon wafer is "floating" inside the hockey puk package, the pressure applied to each side

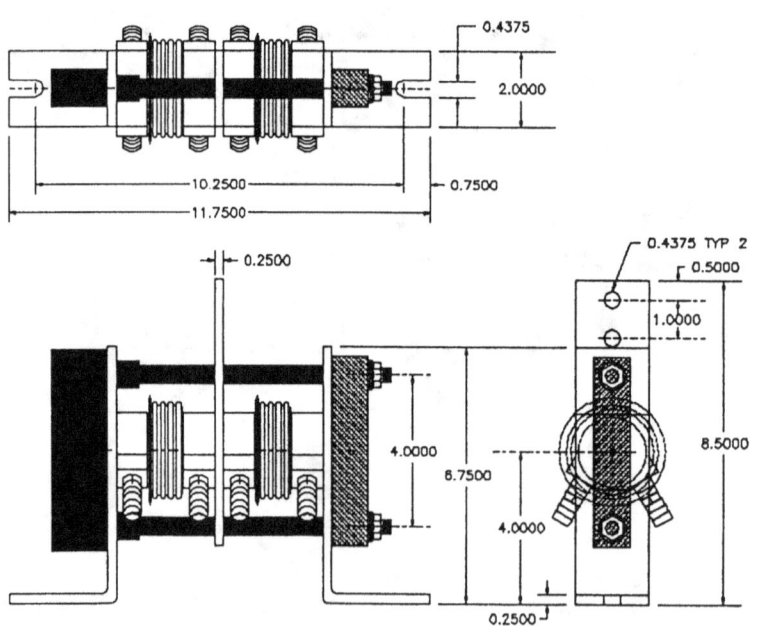

Figure 17 – Dual SCR Water-Cooled Heatsink Assembly
(drawing courtesy of Value Engineered Products)

of the device not only holds it in place but also ensures a good electrical connection internal to the SCR and externally between the SCR and heatsink. Care must be taken to ensure the pole faces of the SCR are parallel after clamping is complete. If the pole faces are not parallel, then current will be channeled through a small portion of the silicon wafer causing the SCR to run hot and fail prematurely.

Another key ingredient of a good SCR/Heatsink thermal design is the use of a proper heatsink compound. The heatsink compound not only helps ensure the proper transfer of heat from the SCR to the heatsink, but in some cases, such as with a hockey puk SCR, it also helps maintain a good electrical connection.

Since most heatsink compounds are similar to petroleum jelly in consistency only with higher temperature and conductivity ratings, it is possible to apply too much to the surfaces. You should only apply a very thin layer of the compound to the surface of the SCR. Too much will prevent proper tightening or become a high resistance thermal connection; and, in the case of the case of a hockey puk, a poor electrical connection as well.

SUMMARY

Although it's been more than 40 years since its initial design, the SCR is still a popular device for controlling large amounts of electrical power. The many voltage and current ratings, different mechanical packages and large number of manufacturers are a few of the reasons the SCR has survived so many years and is still one of the leaders for solid-state switching of electrical power.

Chapter 5

FIRING CIRCUIT CONTROL METHODS

GENERAL

There are two basic types of control methods used in SCR Power
Controls - Zero-Firing & Phase-Firing and each has it's own
specific advantages and disadvantages. Each application should
be reviewed based upon: type of heating element, whether or not
the load is transformer coupled, possibility of shorts in the
load, load temperature stability requirements and Power-Factor
and harmonic current and RFI concerns. Each of these helps
determine the proper control method to use.

ZERO-FIRING

It seems that nearly every SCR Power Control manufacturer has its own terminology for Zero-Firing of SCRs such as: time proportioning, zero crossing or even distributive control. However, no matter what terminology is used, it is basically the same. The term Zero-Firing comes from the fact that the SCRs are switched on at zero current or the zero crossing point on a sine wave as shown in Figure 18. Note that the SCRs are always turned on at zero crossing and are full sine waves. In Figure 18, the incoming power is shown on the left and the SCR output on the right. Since the current is being switched on at zero current, no fast high rising waveforms are present. Therefore there are basically no harmonics or RFI generated. In addition, since the SCR is conducting for the entire sine wave, the power factor will be near unity.

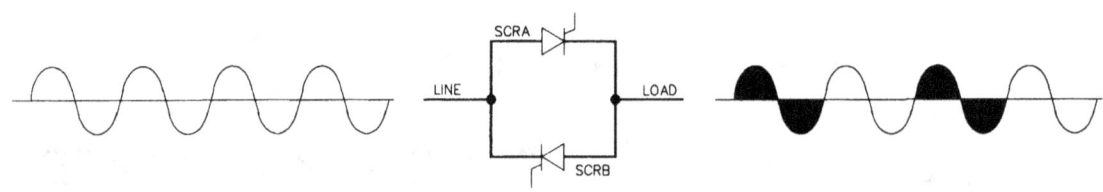

Figure 18 - Zero-Firing waveforms
(The dark areas indicate SCRs turned on)

To complicate matters, there are two types of Zero-Firing: Variable-Time-Base (VTB) and Fixed-Time-Base (FTB). Once again, each of these has its own advantages. However, most users agree the variable-time-base advantages outweigh those of the fixed-time- base and are the most popular.

Variable-Time-Base (VTB)

As in any Zero-Firing control mode, the VTB unit will

change or vary the sum of the number of ON/OFF AC cycles to meet the demands of the load. For example, at 50% command the SCRs would be ON for 3 cycles and OFF for 3 cycles for a total time base of 6 cycles. At 75% command the SCRs will be ON for 9 cycles and OFF for 3 cycles for a total of 12 cycles. At 90% the SCRs will be ON for 27 cycles and OFF for 3 cycles for a total of 30 cycles. As you can see, the total time base continuously changes to optimize the ON time. By minimizing the OFF time, heating elements are kept at a more constant temperature; which, in turn, minimizes thermal shock and improves life expectancy.

Different manufacturers use different total time base lengths. However, the longer the Variable-Time-Base is, the better the overall control resolution.

The Variable-Time-Base minimizes thermal shock, helps prevent light flicker and extends the life of other electrical equipment such as fuses and transformers by not thermally shocking and/or mechanically stressing them. In addition, the power delivered to the load is more constant which can result in improved processes.

Fixed-Time-Base (FTB)

Although similar to the Variable-Time-Base firing method, the Fixed-Time-Base has one big difference. As you noted with the VTB, the total number of AC cycles varied from as few as 6 to possibly a hundred or more. With the FTB the total time base remains the same regardless of the operating point.

For example, assume a FTB of 3 seconds. At a 50% command, the SCRs would be ON for 90 cycles and OFF for 90 cycles for a total of 180 AC cycles. For a command of 75%, they would be ON for 135 cycles and OFF for 45, a total of 180 cycles. Again, at 90% the number of ON cycles would be 162 and the number OFF would be 18, again a total of 180.

As you can see, you can never achieve the resolution of the variable time base. The time OFF can be much greater causing thermal cycling or shocking to the load and other electrical components. Also, if you are switching large currents, the pulsing of the ON/OFF cycles can cause flickering of lights in the work place.

PHASE-FIRING

Phase-Firing (sometimes called Phase Angle Firing) gets it name from the fact that the SCRs are turned on at different "phase angles" or at various times within each half cycle of the AC power.

The use of Phase-Firing gives better resolution than Zero-Firing and is normally used for transformer coupled loads, fast responding loads or loads that have large resistance changes with temperature. A tungsten element is a typical example of a fast responding load. Molybdenum Disilicide is a load that has a large resistance change with temperature.

A look at Figure 19 gives you a good idea of what the chopped waveforms Phase-Firing creates. It is this steep, high rising waveform that causes RFI. The waveform on the right indicates how only a portion of the waveform is delivered to the load creating a low Power-Factor. Power-Factor is a utilization factor. You pay for the complete sinewave but only use a portion.

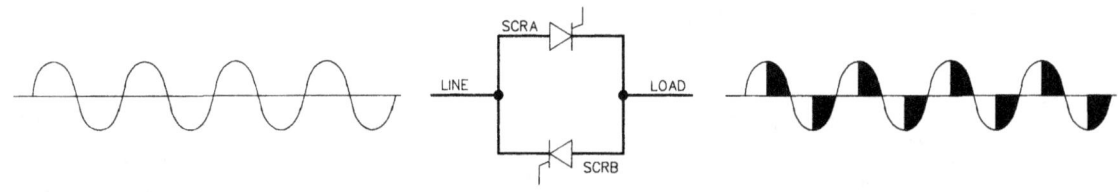

Figure 19 - Phase-Firing Waveform
(Note the chopped waveforms on the SCR Output)

Phase-Firing is typically more expensive than Zero-Firing due to the complexity of the firing circuit. Also, on three phase units two leg control is not available.

In the early development years of the SCR Power Control, two different types of three phase/three leg control circuit configurations were used: the Hybrid (3 SCRs and 3 Diodes) and the Three Phase Switch (6 SCRs).

Hybrid Circuit

The Hybrid circuit, as the name implies, is made up of a Diode and an SCR connected back to back in each phase. This was used most of the time in the early years of SCR Power Controls because the price differential between SCRs and Diodes was significant. That difference is not as great today and the disadvantages of the Hybrid Circuit are more of a problem. Refer to Figure 20.

The greatest disadvantage was a more distorted waveform that caused higher harmonics content; and, since the waveform was not symmetrical, use on unbalance loads required significant over- sizing of the SCR Power Controller. In addition, this circuit could not be used on "4 wire" connected loads. However, it could be used on either Phase-Fired or Zero-Fired applications.

Three-Phase Switch

The three-phase switch is simply 3 AC switches as discussed earlier in Chapter 5. This circuit configuration provides the most symmetrical waveforms that help to minimize harmonics and allows for operation on unbalanced and/or "4 wire" loads. This circuit is used almost exclusively now. See Figure 21.

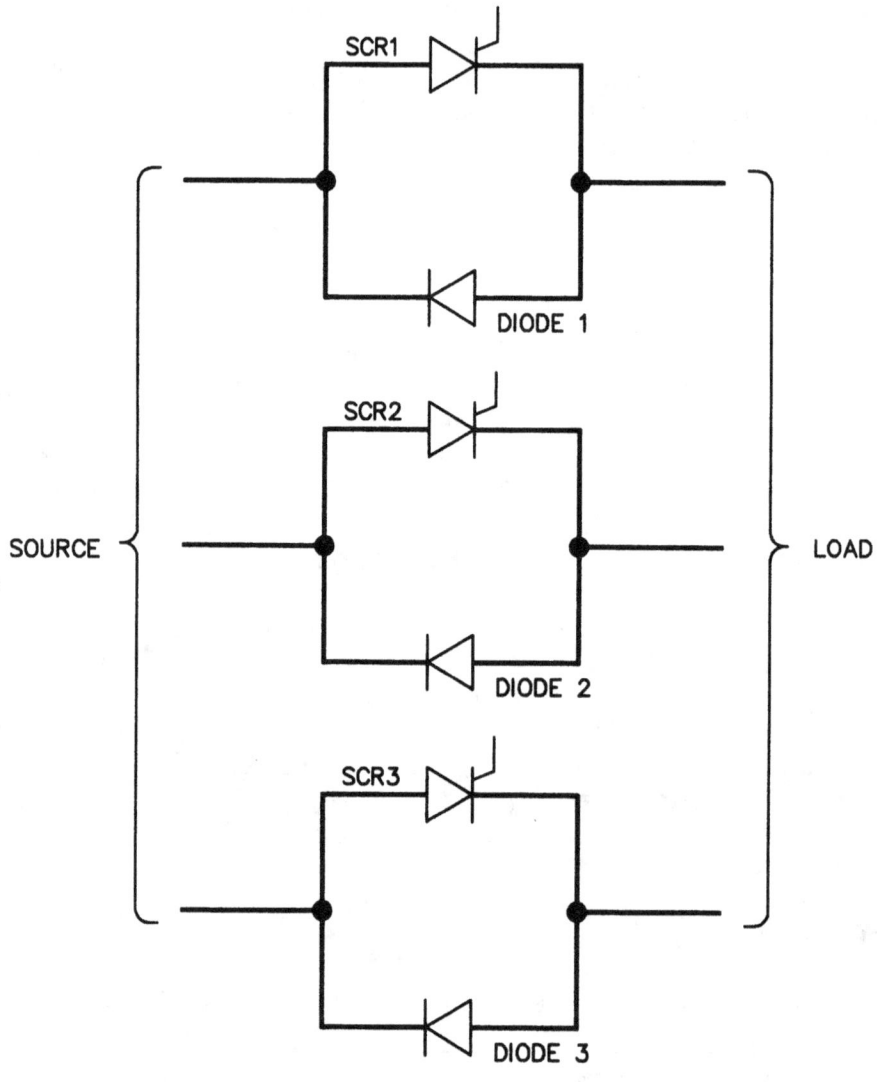

Figure 20 - Schematic of a 3 Phase Hybrid Circuit

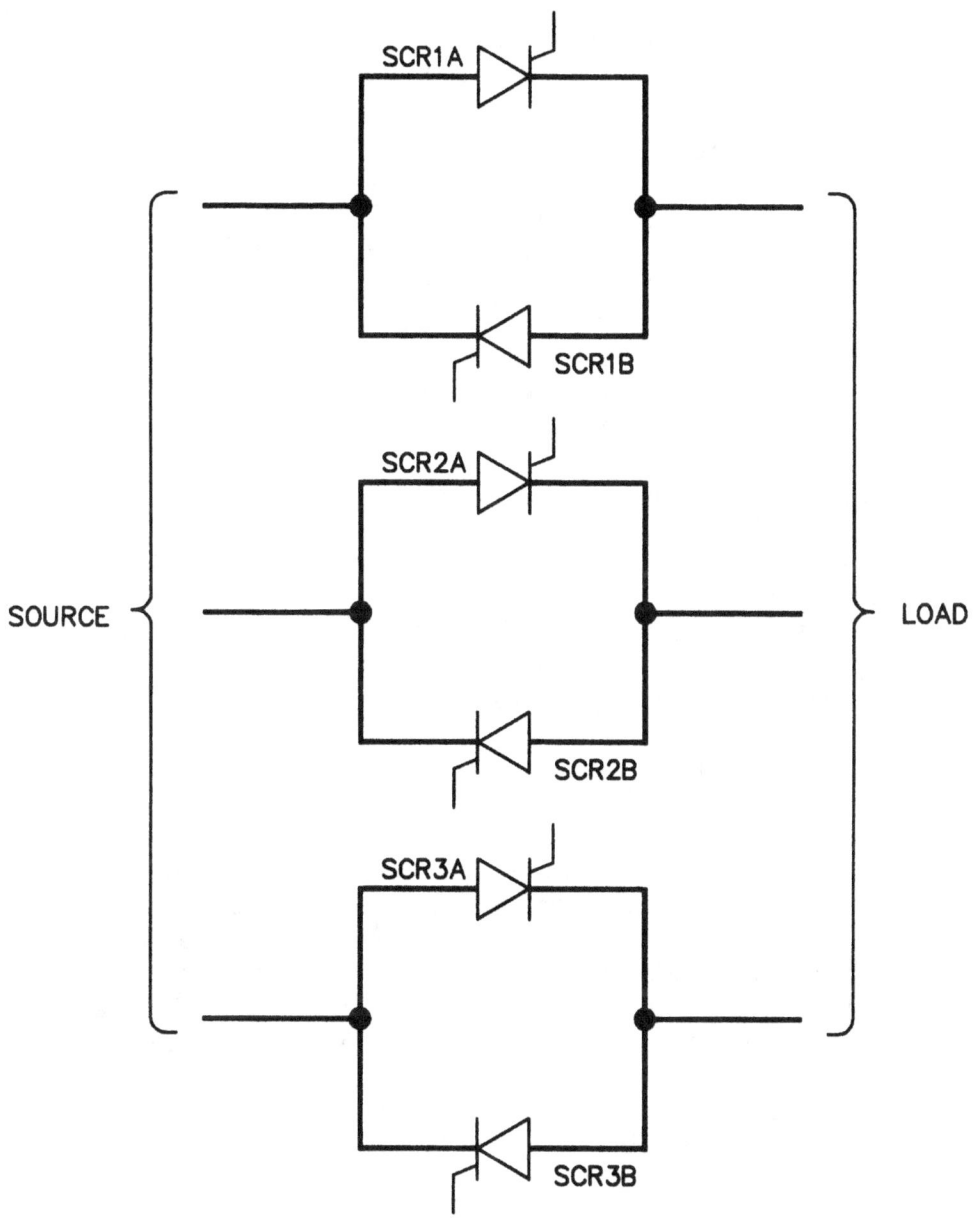

Figure 21 - Schematic of a Three-Phase Switch

Another important requirement for Phase-Firing is the start-up ramp or the Soft-Start, as it is commonly referred to.

Soft-Start

This circuit provides a slow walk up of power when the command signal is calling for power. If the Soft-Start time period is not correct, the output power will increase too rapidly which can cause trouble depending upon the type of load. In the case of a transformer coupled load, core saturation can occur if power is applied too quickly to the primary. If controlling a load with a high resistance change such as Molybdenum Disilicide, an overcurrent can occur because the element's resistance is nearly a short circuit until its temperature rises. The Soft-Start is usually an R-C time constant which allows for a ramp of approximately 10 to 15 AC cycles. This is sufficient time to allow driving the primary of a transformer safely without core saturation or controlling a Molybdenum Disilicide load while using RMS Current Regulation.

Another important part of the Soft-Start circuit is the Soft- Start reset. This circuit serves as a safety net. If the power is turned off intentionally or if the power line "blinks", the Soft-Start must reset to zero and then ramp up if the command signal is calling for power. If the Soft-Start is not reset during a power outage, then when power is re-applied the Soft-Start ramp will have already timed out and surge currents will occur due to transformer core saturation. In the case of a Molybdenum Disilicide element, if it has cooled off, then its resistance will be very low and an over current surge can happen. The reset circuit must be very fast so reset occurs before unwanted short duration power outages cause problems.

SUMMARY

Although there are two types of Firing Circuit control methods, they have many similarities. Simply put, Zero-Fired should be used on non-inductive loads and Phase-Firing on inductive and transformer coupled loads or loads with large resistance changes. Zero-Firing has near unity power factor (PF) and minimal harmonics and RFI. Phase-Firing can cause poor Power-Factor and generate harmonics and RFI. Zero-Firing is usually the better choice whenever the application permits.

NOTES CHAPTER 5

Chapter 6

TYPES of FEEDBACK

GENERAL

When thinking of feedback with relative to SCR Power Controls, it can mean any of the following:

* A means to linearize the output power level as compared to the input command signal for a given parameter such as RMS Voltage, Current or Power Regulation.

* A means to compensate for line voltage fluctuations.

* A means to limit certain parameters such as RMS Current.

* A means to shut-down the SCRs in case of excessive current.

Many SCR Power Controls are not limited to only one type of feed-back but can use as many as two or three on one application. For the purposes of this book, I will limit the meaning of feedback to only those previously mentioned.

Many SCR Power Controls have numerous options available to regulate voltage, current, etc. However, each of these have an additional cost associated with them and for many applications simply are not required. In these cases, most manufacturers have designed an internal type of feedback to help regulate the "SCR's" output. In some cases, it may be duty cycle linear for a Zero-Fired unit or, in the case of a Phase-Fired unit, it may be phase-angle linear.

The method I have used over the years has been voltage squaring. Basically this circuit performs two functions for the user. First, since you are trying to control power, why not linearize the power output? This circuit assumes a fixed load resistance and, in turn, linearizes the output power to the input command signal level. Keep in mind this is not true power regulation since it is not actually measuring load voltage and current and computing power. It is only a simulation. The second function is line voltage compensation.

LINE VOLTAGE COMPENSATION

Line Voltage Compensation is the simplest form of feedback but still has a very useful purpose. This feedback circuit monitors the line voltage and adjusts the controller's output level up or down as required to maintain a constant voltage output regardless of line voltage fluctuations. Typical specifications are +/- 3% output change for a +/- 10% line voltage change which is reasonable for the little cost associated with this circuit. This circuit can be used on both Zero and Phase-Fired equipment.

RMS CURRENT LIMIT

Sometimes referred to as a current clamp, this circuit acts as a safety net. It monitors the load current and limits or clamps it to a preset value regardless of line voltage or load resistance changes. Typical circuits hold the output RMS current within 1% with +/- 10% line voltage changes and resistance changes of up to five times. The current limit is normally a slow device and is used in conjunction with an over current trip circuit.

Current Limits are especially useful to help prevent large current surges during start-up due to loads that have low resistance when cold. Current Limit is used only on Phase-Fired equipment.

OVERCURRENT TRIP

The Overcurrent Trip is a fast acting electronic circuit designed to "shut off" SCRs in case of a current overload. Some manufacturers refer to this circuit as an "electronic fuse" but do not be fooled. A fuse is still the best answer for short circuit protection. The reason is simple. Since you cannot turn off an SCR until the current through it is zero, the best you can do electronically is to shut down within one-half cycle. Semiconductor fuses are sub cycle and can clear anytime during a half-cycle.

This circuit is still valuable since it provides an extra level of protection and can shut down prior to clearing a fuse. Like a fuse, it should shut down based upon peak currents and not the RMS value.

Overcurrent Trip circuits need reset after tripping. This is usually accomplished by either pushing a reset button or by removing power to the circuit. In addition, most manufacturers include a relay output for alarm or indication purposes.

RMS VOLTAGE REGULATION

This circuit regulates the RMS voltage applied to the load regardless of line voltage fluctuations or load resistance changes. Typical specifications are +/- 1% output change for a +/- 10% change in line voltage or load resistance.

47

Since it is possible to overcurrent the SCR Power Controller, a RMS Current Limit is recommended in conjunction with the RMS Voltage Regulator.

RMS CURRENT REGULATION

This circuit is virtually identical to the RMS Line Voltage Regulator except that it is monitoring load current via a current transformer (CT). Specifications are typically +/- 1% output change for a +/- 10% line voltage change or load resistance change up to 50% or more.

Because Phase-Firing distorts the waveform, a true RMS Current Regulator or Limit is required. Some SCR Power Control manufacturers use average current reacting Current Regulators and Limits because the cost is less to build. Do not be fooled, specify true RMS Current options.

TRUE POWER REGULATION

True Power Regulation is similar to voltage and current regulation except this method measures both the load voltage and current, multiplies them and uses that product for feedback. The multiplier Integrated Circuit is normally referred to as a four-quadrant multiplier. It takes into account the phase angle and phase shift of the output waveform to produce a signal proportional to true power. Once again typical specifications are +/- 1% output change for a +/- 10% line voltage change and a +/- 50% load resistance change.

Since the possibility to over current the SCR Power Controller exists, always use a RMS Current Limit in conjunction with a True Power Regulator.

SUMMARY

There are several types of feedback available for use with SCR

Power Controls which give the engineer different solutions to his/her power control problems. Each one has its own features and it is simply a matter of the application and any cost restraints that may force the engineer to eliminate the extra circuits.

NOTES CHAPTER 6

Chapter 7

POWER-FACTOR, HARMONICS & RFI

GENERAL

The use of Phase-Fired SCR Power Controls can lower Power-Factor, increase Harmonics and possibly generate Radio Frequency Interference (RFI). A poor Power-Factor can result in the user paying more for electrical energy that is undesirable. With the slow adaptation of IEEE standard 519 throughout industry, harmonics and RFI have become a common concern. To my knowledge, very few problems have occurred in actual applications.

POWER-FACTOR

Power Factor is a utilization factor of the power being consumed from the source. It is expressed as a decimal with unity (1.0) being perfect. Anything less than 1.0 (0.65 for example) is considered poor. Disregarding losses, unity Power-Factor indicates that all power being consumed is being delivered to the load. It is also expressed as leading or lagging indicating whether the current is leading or lagging the voltage due to inductance or capacitance in the circuit.

Since most utility wattmeters indicate Power-Factor (P.F.) as near "unity" for zero-fired equipment, all the remaining discussions will relate to phase-fired SCR Power Controls. Also for simplicity's sake, discussions are limited to resistive loads. Since balanced three-phase circuits have the same power factor as single-phase circuits, both will be treated as equal.

Power Factor is the ratio of the true power (in watts) divided by the apparent power (volt-amperes).

$$P.F. = \frac{Watts}{Volt\text{-}Amperes}$$

Since the load and source currents are equal, the power factor in a resistive load is simply the ratio of the load RMS voltage to the source RMS voltage. For example, to calculate the power factor of a circuit with 340 Volts RMS applied to the load from a 480 Volt RMS source, simply divide the load voltage by the source voltage.

$$P.F. = \frac{340}{480}$$

P.F. = 0.71

IMPROVING POWER FACTOR

The first defense against poor power factor is to always use a zero-fired power control if possible. The addition of a tapped transformer to the Phase-Fired SCR Power Control circuit will help. The addition of Power-Factor capacitors can also power utilization.

If using the tapped transformer method, each tap should be selected as a percentage of the next higher tap. For example, if the user wants to maintain a 0.9 P.F. then tap 1 would be 90% of tap 2, tap 2 would be 90% of tap 3, etc. The user would then operate the power control system on the lowest voltage tap possible to maintain temperature. The higher the percentage of full voltage they operate on any tap, the higher the power factor. If the process requires a wide range of voltages, a tap switch will enable the operator to change taps easily and quickly.

The use of P.F. Capacitors (rated in kVAR) can do a very good job of improving Power-Factor. However the kVAR requirements change as the load characteristics change. P.F. Capacitors are typically selected with the most typical operating point in mind. I usually recommend capacitors be installed for the entire plant load (not just the power control load). By doing this, the capacitor requirements will change less with load variations and the result is improved power factor over a wider range of control.

There is one important issue to remember when specifying Power-Factor Capacitors. The addition of capacitors into a plant power distribution system can react with other equipment connected to the same electrical source producing a number of problems with the SCR Power Control and other connected equipment. It is important to have a complete plant electrical system study by individuals skilled in the application of Power-Factor Capacitors.

HARMONICS AND RFI

Harmonics are generated when electrical energy is switched. Any electrical controlling device such as SCR Power Controls, mechanical contactors, saturable core reactors or most other power controlling devices can generate harmonics.

Harmonics are expressed as a multiple of the fundamental frequency (60 Hz for example). For example, a harmonic with a frequency five times the fundamental would be referred to as the 5th harmonic. The magnitude or energy content of each harmonic is lower as the multiple increases. For example, the 12th harmonic would have lower energy content than the 3rd harmonic.

Radio Frequency Interference (RFI) is caused by the abrupt switching action of a controlling device. In the case of an SCR Power Control, RFI is greatest when the SCR conduction angle is at 90 degrees because the waveform has the highest peak switching action at that point. This is the reason Zero-Fired Power Controllers produce minimal harmonics they are switching at zero current.

Note in Figure 22 the output waveforms are clean, no chopping, no high rising waveforms. The Zero-Fired waveforms are virtually identical to the incoming power allowing complete utilization of the waveform and providing a near unity Power-Factor.

As you can see in Figure 23, Phase-Firing creates chopped waveforms that create Radio Frequency interference. Also, only a small portion of the incoming waveform is used by the load causing a poor utilization factor or poor Power-Factor.

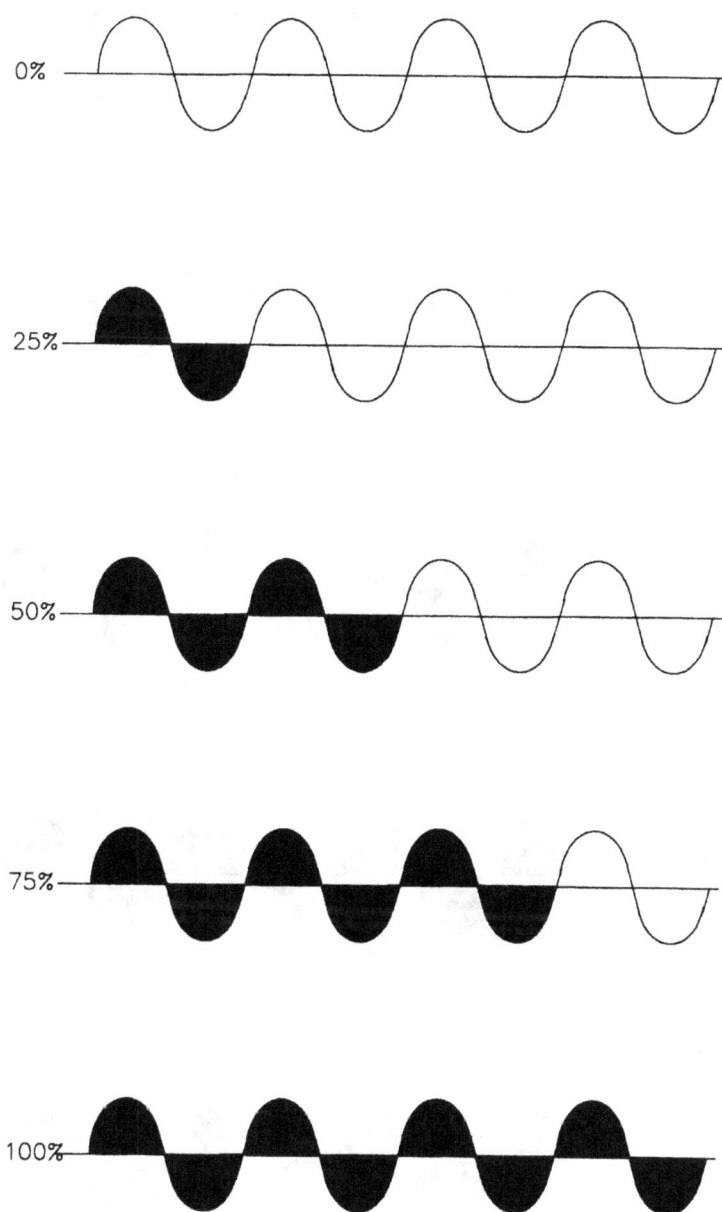

Figure 22 - Zero-Fired SCR Waveforms at Various Output Levels

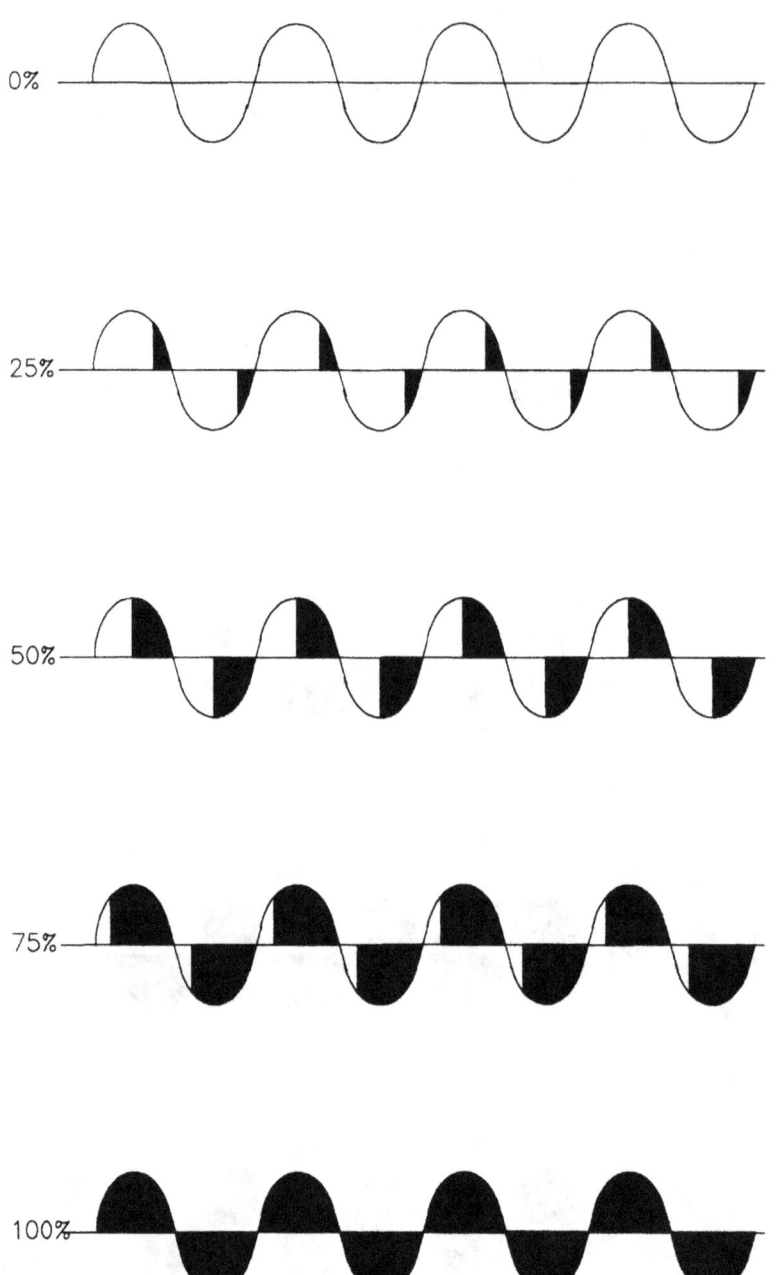

Figure 23 - Phase-Fired SCR Waveforms at Various Output Levels

The chart in Figure 24 indicates the magnitude of the odd harmonics for 90°. The even harmonics are not shown since their magnitude is near zero. The chart also shows the frequency of the harmonic that is simply the number of the harmonic times the base frequency. As you can see, as the number of the harmonic increases the magnitude decreases.

Harmonic	Frequency	Magnitude (%)
1	60	59.7
3	180	31.8
5	300	10.6
13	780	4.6

Figure 24 - Odd Harmonics at 90° Conduction
(The Frequencies are derived from a base frequency of 60 Hz.)

SUMMARY

Poor Power-Factor is a concern since it can increase electricity cost. Harmonics and RFI are partially the result of poor power factor caused by low conduction angles of Phase-Fired SCR Power Controls. However, there are methods to deal with each of these. The eventual adaptation of the IEEE 519 standard will make these more of an issue. The use of Zero-Fired SCR Power Controls can help eliminate this problem.

Chapter 8

LOAD CONSIDERATIONS

GENERAL

One of the first evaluations an engineer must make when
selecting an SCR Power Controller is which type of load it will
be controlling. There are five basic types of loads that require
understanding of their resistance characteristics. For example,
one load may have a large resistance change with temperature
while another has very little or one load may be very fast
responding while another has a large resistance change with
time. One needs to understand each of these fully to select an
SCR Power Control.

BASIC TYPES OF LOADS

The five basic types of loads are the following:

Resistive - Defined as any resistive load element with a resistance change of less than 10%. These are the easiest types to apply because there are no special features to consider. Purely resistive loads are typically used on lower temperature applications. Iron Chromium and Nickel Chromium are two examples of resistive type elements. Although Phase-Firing will work, Zero-Firing is the best choice. Generally, no feedback options are required.

Resistance Changes with Time - The most popular element with this characteristic is Silicon Carbide. The resistance starts high when new and decreases with age then again increases as aging progresses. Review Figure 25 for typical resistance characteristics.

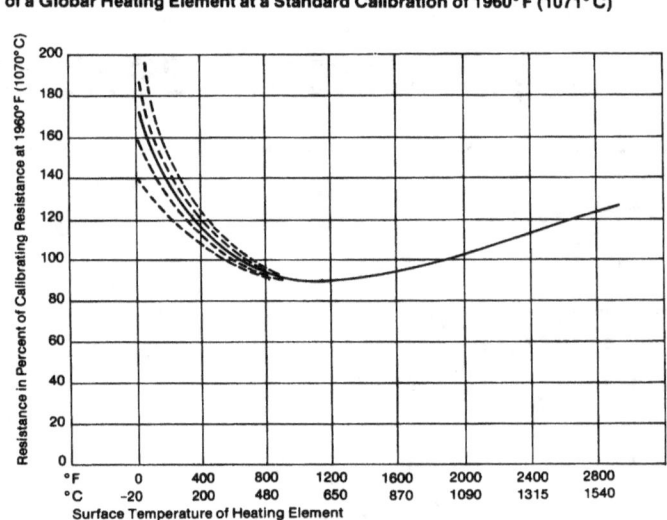

Typical Resistance Temperature Characteristic of a Globar Heating Element at a Standard Calibration of 1960°F (1071°C)

Figure 25 - Resistance Curve for Silicon Carbide
(chart courtesy of Cesiwid Corp.)

Silicon Carbide can be controlled using either Phase-Firing or Zero-Firing. Phase-Firing is the most popular with three methods available.

1. The simplest method is simply using a Phase-Fired SCR Power Controller with Current Limit. In this case the user simply turns the Power Controller on, allows it to operate at full voltage and as the resistance decreases, the Current Limit prevents an over current situation. The biggest drawback is the Power Factor can be very poor during the low resistance stage. This has the lowest equipment cost.

2. The next method is adding Power Regulation to the SCR Power Control. The Power Regulation option will monitor the load voltage and current and provide a feedback to linearize the Power Control's output. Unlike method 1, which only controls the load voltage, this method maintains a constant power over the entire load resistance curve. Similar to the previous method, it presents a Power Factor problem. This produces medium equipment cost.

3. The third and most commonly used method is including a tapped transformer to either method one or two. This allows the user to maintain a high Power Factor during the entire life of the Silicon Carbide element. It does have the highest equipment cost; however, over the long run it may cost less due to improved energy usage.

The most common tap arrangement on the transformer is six evenly spaced taps. The voltage range is typically 2:1 with tap 1 as the lowest voltage. Taps 2 through 6 are full kva with tap 1 a reduced kva. This tap is commonly referred to as the "idling tap" and is typically used to maintain a lower temperature in the furnace on weekends or during downtime.

In any of the three Phase-Firing methods or if using a Zero- Fired controller, the engineer must size the SCR Power Controller's current rating based upon the heating element's low resistance stage or "Valley" as it is commonly called.

Resistance Changes with Temperature - These elements present more of a control problem since most appear as a short circuit when cold. In this case, Phase-Firing is required because very fine control is required while the element is cold. Molybdenum, Tungsten, Stainless 304, Quartz Discilicide and Graphite are typical examples. In each element, the resistance increases as temperature increases. However, if cold, power must be applied slowly. Always include a Current Limit to "help" hold the current to a safe level during start-up. If the power is not applied slowly, extremely high surge currents will occur which can cause damage to the SCR Power Controller, the heating element or possibly other equipment.

Fast Responding Loads - There are many different types from Infrared Lamps or Low Mass Heaters to Arc Processes. Since high surge currents are common and high resolution is normally needed, Phase-Firing is required.

Transformer Coupled Loads - Any of the previously mentioned types of loads can be transformer coupled. In general, due to the high cost of the transformer, it would only be used if the load element had a low voltage requirement that would prevent it from operating directly off the incoming power.

VOLTAGE/CURRENT CALCULATIONS

The following formulas are simple and easy to remember. They can be used over and over when selecting and sizing SCR Power

Controllers.

Single-Phase Connected Loads - Simply use "ohm's law" to determine the unknown parameter. To use Figure 26 as an aid, simply put your finger over the unknown parameter and use the remaining formula to solve the problem.

E = VOLTAGE
I = CURRENT
R = RESISTANCE
P = POWER

Figure 26 - Ohm's Law Charts

One simple tip to remember when calculating 3 phase voltages, currents or power (kva) is the "square-root of three". It is easy to remember - associate the 3 in three-phase to the 3 in the "square-root of three" and you will never forget how to calculate three-phase currents and voltages.

Three-Phase Delta Connected Loads - Refer to Figure 27 for these calculations. As you can see, in a Delta configuration, the heating element has the full line voltage applied to it. This is a major distinction between Delta and Wye loads. You should also note the circuit schematic resembles the Greek letter.

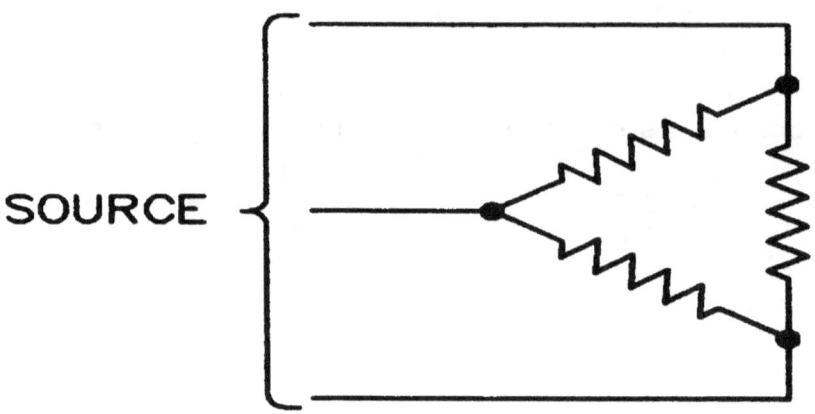

Figure 27 - Three-Phase Delta Connected Load

Three-Phase Wye Connected Loads - Refer to Figure 28 for these calculations. Once again, the "square-root of three" serves a useful purpose in calculating the individual heating element's voltage. Also note, the schematic diagram is shaped in a Wye.

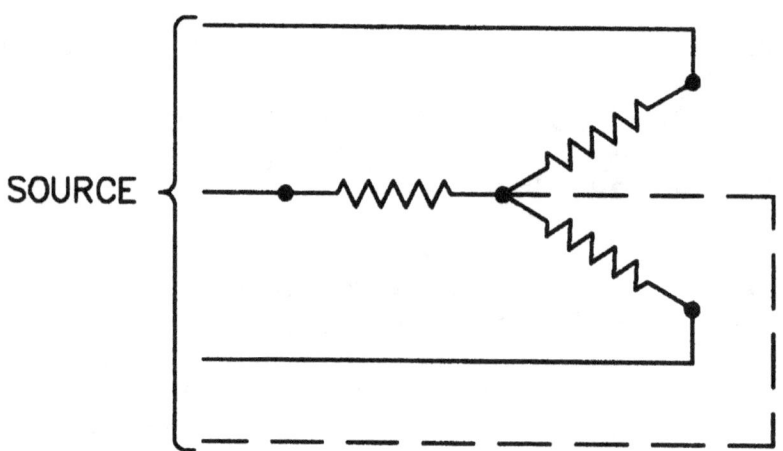

Figure 28 - Three-Phase Wye Connected Load

This formula can also be used for calculating voltages, currents

64

or kVA requirements for transformers.

SUMMARY

There are many types of loads made of various materials that have different resistance characteristics from each other. Special care is required when selecting an SCR Power Controller considering not only the voltage and current rating, but also the type of firing and the options all based upon the type of heating element.

Single and Three-Phase voltage, current and power calculations are simple. The "square-root of three" is important to performing any three-phase calculation.

Chapter 9

COOLING CONSIDERATIONS

GENERAL

Keeping semiconductors cool is the key to the overall life expectancy of any solidstate device. There are many ways of keeping semiconductors cool. Both passive and active methods are available depending upon the application or installation.

PASSIVE METHODS OF COOLING

In general, passive cooling (relying upon natural convection to remove the heat) is usually limited to lower current ratings - usually less than 200 amps. However, some manufacturers of SCR

Power Controls have pushed the envelope by either simply increasing the discrete SCR rating, increasing the heatsink size and lowering the current rating of the SCR Power Control or using duty cycle ratings of less than 100% to achieve higher ratings. These are not the preferred methods.

The first method of increasing the discrete SCR's rating and the heatsink size and lowering the current rating of the SCR Power Control ultimately increases the Power Controller's size which is undesirable. In addition, the increased cost of this method far out weighs the cost of active cooling (a cooling fan for example) and usually results in a higher selling price.

Figure 29 - Multi-Zone Power Control System
(photo courtesy of HDR Power Systems, Inc.)

Figure 29 is a good example of a multi-zone Power Control System. In this case, the use of passive cooling would definitely increase the overall size of the system and increase the cost.

The use of a duty-cycle rating on a Power Controller is a disaster waiting to happen. Granted, most applications have a typical operating point less than 100% of full output, for example, 80%. However, during start-up the SCR Power Control will be operating at full load for long periods of time. During this time, if the Power Controller has a duty-cycle rating of less than 100%, the discrete SCRs will be operating at temperatures higher than designed for, which in turn will decrease the life expectancy. Since the mass of an SCR is small, its internal temperature reacts to current changes very quickly.

In the example of an 80% duty-cycle rated unit controlling a furnace with a typical operating point of approximately 80% of full load during normal operation, the SCR Power Controller will be operating within its rating. However, during the start-up period which can last hours or days, the Power Controller will be operating at 25% over its steady state rating. Although failure may not be immediate, degradation of the discrete SCR occurs each time it is operated above its temperature rating.

ACTIVE METHODS OF COOLING

The two most widely used methods of active cooling are forced air and water. Both are efficient ways of removing heat from the semiconductor.

Forced Air Cooling - The simple addition of a cooling fan to a heatsink assembly will lower the semiconductor's temperature substantially therefore increasing the life expectancy. The cost of adding a fan is small and the excuse that fans are unreliable is no longer valid. Fan manufacturers have worked hard at increasing fan performance and reliability.

One drawback to fans is the electrical power requirements. Since nearly all control panels have internal control power, the small power requirements of the fan is normally not a problem. The other drawback if it is installed in a dirty environment is that the fan can force dirt and dust onto the heatsinks, which can decrease the heatsink's cooling efficiency. The addition of air filters is inexpensive and reliable. My experience says users of this type of equipment understand the need for clean air and will change out the air filter whenever needed. If the installation is clean, no filter is required.

Water Cooling - This is the most efficient method of cooling semiconductors. The obvious drawback is that if

Figure 30 - Typical Closed Loop Water Cooling System
(photo courtesy of Water Saver Systems, Inc)

the process being controlled does not use water or there is no plant cooling water, this is not a viable choice. Water requirements are small, typically only 1 to 1 1/2 Gallons per Minute per heatsink assembly.

CAUTION!!!!!

Depending upon which type of heatsink is used, the cooling water can be electrically hot! Always use "non-conducting" rubber hose to connect the cooling water to the heatsink. Depending upon the water quality, a hose length of about 18 inches for a 480 Vac system is required.

Water quality is important. Besides being free of dirt and debris, the following water specifications should be used as a general guideline. By controlling each of these parameters, the cooling water will stay clean. Filtering the cooling water is always a good idea.

Cooling Water Characteristics (guideline)

* Total Hardness of CaCO3 - 100 ppm

* Total Dissolved Solids - 200 ppm

* Conductivity - 300 micro mhos/cm

* pH - 7.0 to 8.0

* Suspended Solids - 10 ppm

In addition to this, the maximum water inlet temperature should not exceed 102°F (40°C) and the minimum temperature should be maintained above the dew point to avoid condensation. The water flow and pressure should be as specified for the equipment.

SUMMARY

There are two common methods of cooling semiconductors: passive cooling such as convection air cooling and active cooling which includes both forced air and water. Each of these has both strong and weak points. The user needs to evaluate the process and the environment to determine which method is best.

Chapter 10

ENCLOSURE SELECTION

GENERAL

Most all applications for SCR Power Controllers require some type of enclosure. This chapter is intended to assist in properly selecting an enclosure, taking into consideration the environment and the SCR Power Control.

> Note: This chapter is provided courtesy of HDR Power Systems and is essentially a copy of HDR Application Note - 1002. I wrote Application Note - 1002 a few years back in preparation for this book.

SELECTION CRITERIA

The first step in selecting an enclosure is to evaluate the environment. Answers to the following eight questions are essential to making the proper enclosure selection.

1. What is the maximum ambient temperature where the enclosure will be installed?

2. Will the enclosure see any reflected heat from any nearby heat source?

3. Is there dust, water, oil or other contaminants that may affect the overall operation of the equipment?

4. Is the user installing any additional equipment in the enclosure?

5. Are both sides, top, bottom, front and back exposed to the air for heat sinking purposes?

6. Are there any objections to forced-air cooling?

7. Is cooling water available if a sealed enclosure is required?

8. Are there any size restrictions?

TYPES OF ENCLOSURES

Once the above questions are answered, you can select an enclosure type for the following list:

NEMA 1 General Purpose - used indoors as a safeguard for personal safety, normally where special conditions do not exist. NEMA 1 enclosures may be ventilated.

NEMA 2 Drip Proof - used indoors where falling, non-corrosive liquids and falling dirt may damage electrical/electronic components.

NEMA 3 Dust Tight - for outdoor use where protection from wind-blown dust and water is required.

NEMA 3R Rain Tight - same as NEMA 3 except NEMA 3R enclosures provide equipment protection from the rain. They are not dust tight or sleet-resistant.

NEMA 4 Watertight and Dust Tight - for indoor or outdoor use where splashing or seeping water, falling or hose-directed water and severe external condensation are threats to the electrical/electronic equipment.

NEMA 12 Industrial Use, Dust Tight and Drip Tight - used indoors for protection from fibers, dust and dirt, light splashing, seepage, dripping and external condensation of non-corrosive liquids.

Note: There are other types of NEMA ENCLOSURES for special Applications beyond the scope this chapter.

SIZING THE ENCLOSURE

After selecting the enclosure, you can select the size. This requires knowing the following:

1. The ambient temperature at the installation
2. The SCR Power Control's overall dimensions
3. Additional equipment installed in the enclosure
4. The maximum allowable equipment operating temperature

Space is always at a premium, so initially select the smallest enclosure that will hold the equipment while allowing proper clearance for the mechanical, electrical and thermal considerations.

Calculating Power Loss

The next step is to calculate the power losses into the enclosure for each piece of equipment installed. If there is more than one of the same type, multiply the losses by the quantity. When calculating power losses for any SCR Power Controller, a conservative estimate is given in the following formula:

Power Loss = Max Current x 1.5 x Number of controlled Legs

Enclosure Heat Sinking Capabilities

After calculating the power losses, determine the heat sinking capability of the enclosure. Find the square inches of surface area required for the maximum temperature rise allowed in Figure 31. Subtract the maximum ambient temperature from the equipment maximum allowable temperature rating to find the maximum temperature rise allowed. The surface area required is the product of the total losses in watts multiplied by the surface area required per watt from Figure 31.

Now we have determined the type of enclosure, the estimated size and the surface area. Next, one needs to verify the thermal size.

In the following example, assumed are the following:

1. Only one Power Control will be mounted in the enclosure (an HDR ZF1-480-120);

2. The external ambient conditions will be 30°C or less;

3. There are no restrictions on enclosure size;

4. The enclosure will be mounted on a wall which allows the top, bottom, two sides and the front for heat dissipation;

5. A dust-tight and drip-tight enclosure is desired and

6. The actual full load current is 110 amperes.

Example #1

Based on the above assumptions, the procedure for selecting the appropriate enclosure is as follows:

1. Select the enclosure type from the list.

    The correct enclosure is NEMA 12.

2. Select the physical dimensions of the enclosure.

    The Power Control has dimensions of 15"H x 11"W x 10"d. Based on experience, the first pick for an enclosure size will be 10"H x 24"W x 12"d. This allows 7.5" clearance top and bottom, and 6.5" clearance on each side. This is ample room for the Power Controller to fit leaving plenty of room for wiring.

3. Calculate the power losses.

    Loss = 110 x 1.5 x 1

    Loss = 165 watts

4. Calculate the required enclosure surface area.

    The maximum ambient temperature is 30°C and the Power Controller's maximum operating temperature is 50°C; therefore, you can allow for a 20°C temperature rise. From Figure 31, 9.2 square inches of surface area are needed to dissipate each watt of power.

    Required Surface Area = 9.2 x 165

Required Surface Area = 1518 square inches

5. Calculate the available enclosure surface area using the top, bottom, both sides and the front for heat dissipation.

Front:	*30 x 24 =*	*720*
Top:	*24 x 12 =*	*288*
Bottom:	*24 x 12 =*	*288*
Two Sides:	*30 x 12 x 12 =*	*720*
Total =		*2016 square inches*

Since 2016 sq. in. are available and 1518 sq. in. are required, the calculations indicate that the available surface area is greater than the required surface area. This situation indicates a good thermal design and the Power Controller will operate at 100% duty cycle at full current.

FAN COOLING

If the enclosure had not been large enough, select the next size enclosure and rework the calculations. If the application had required a smaller enclosure and allowed forced-air cooling, all of the previous calculations would be performed, in addition the air flow (CFM) requirements would be determined using Figure 32. If a fan is required, and the enclosure size is already determined, calculate the heat dissipation of the enclosure based on Figure 31 using the "Maximum Watts/Sq. In." column. Simply determine the surface area of the enclosure and multiply by the watts/sq. in. number based on the maximum allowable temperature rise.

MAXIMUM WATTS/SQ. IN.	MAXIMUM ALLOWABLE TEMPERATURE RISE	SURFACE AREA REQUIRED SQ.IN./WATT
.054	10°C	18.4
.108	20°C	9.4
.163	30°C	6.1
.218	40°C	4.6
.260	50°C	3.8

Figure 31 - Enclosure Heat Sinking Variables

Next calculate the maximum power generated by the Power Controller and subtract the enclosure heat dissipation. This is the total watts that must be removed by the fan. Refer to Figure 32 to determine the fan size. Sample calculations for this procedure are shown in Example #2.

Example #2

In this second example, the assumptions are 1. one Power Controller will be mounted in the enclosure (an HDR Zf2- 480-225); 2. all other conditions are the same as Example #1 except full load current is 215A. Under these conditions, selecting the proper enclosure is as follows.

1. Select the type of enclosure.

    Again the choice is a NEMA 12.

2.   Determine enclosure dimensions.

The Power Control's dimensions are 15"H x 14"W x 10"D. Using the previous NEMA 12 allows about 7.5" clearance top and bottom and 5" on the sides for mechanical and electrical considerations.

3.   Calculate the power losses.

Loss = 215 x 1.5 x 2
Loss = 645 watts

4.   Calculate the required surface area.

Required surface area = 9.2 x 645
Required surface area = 5934 sq. in.

5.   Calculate the enclosure's surface area.

The calculations are the same as the previous example - 2016 square inches. Since the available is less than the required 5934 square inches, a cooling fan is required.

6.   Determine the watts the enclosure will dissipate (use Figure 31).

Enclosure dissipation = 2016 x .108
Enclosure dissipation = 218 watts

7.   Calculate the watts the fan must remove.

Remaining watts = 645 - 218
Remaining watts = 427 watts

8.   Select the fan.

Using the fan requirements in Figure 32, note that removing 427 watts requires a 40 cfm fan. To extend the applicability of this chart if the power dissipation is higher than the range shown, multiply both the vertical and horizontal numbers by the same factor.

There are many other ways to cool electrical equipment enclosures. Included are air-to-air heat exchangers, direct water cooling, air-to-water heat exchangers, venturi tubes are just a few. These are usually used on equipment installed in dirty installations where it is important to keep the enclosure sealed.

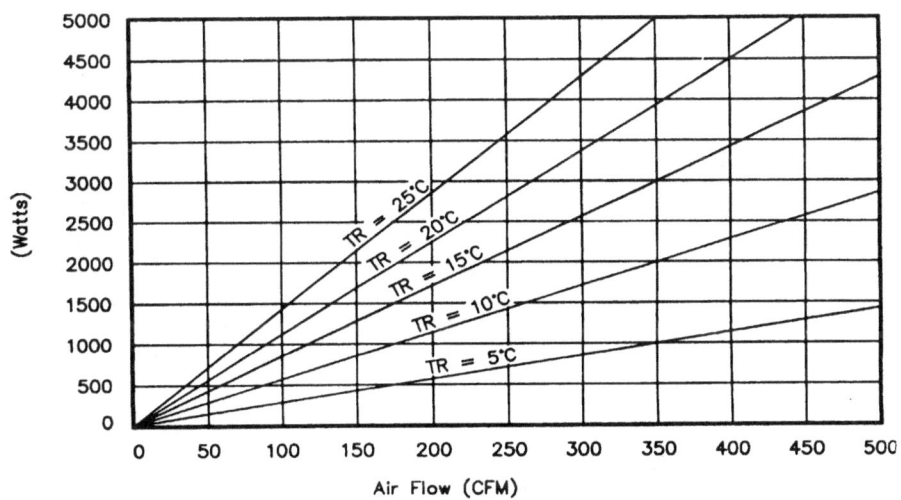

Figure 32 - Fan Requirements Chart

SUMMARY

Sizing an enclosure for a SCR Power Control is not a complicated task. It only takes a few minutes simply by knowing what questions to ask and what formulas to use.

NOTES CHAPTER 10

Chapter 11

MAINTENANCE

GENERAL

Unlike an electro/mechanical contactor, the SCR Power Controller
is solid state and has no moving parts to wear out. Because of
this the routine replacement of contacts is eliminated. It also
does not contain any toxic materials such as mercury in mercury
contactors.

SAFETY

It is important to understand that SCR's are not isolating
devices. All SCR's have a small amount of leakage current when

in the off state. This means the user must have some type of isolating device to completely remove power to the SCR Power Controller during servicing. This can be a Fused or Non-Fused Disconnect Switch, a Circuit Breaker or a Contactor.

MAINTENANCE

As with any electrical power device, the electrical connections should be checked periodically for tightness and corrosion. If corrosion exists, clean the connection and then tighten. If the manufacturer provides torque specifications, then these specifications should be followed. If the information is not provided, then common sense prevails.

As mentioned in previous chapters, the life of the SCR Power Controller is greatly affected by the overall operating temperature of the power semi-conductors and the electronic circuit boards. Periodically check for dirt and grime build-up on the heatsinks. Since the heatsinks dissipate the heat from the SCRs, it is imperative that they are clean. If the Power Control is mounted inside a ventilated enclosure, check the air filters for dirt and ensure that nothing is blocking the air passages.

Smaller current rated Controllers rely entirely upon the heatsink to remove the heat; however, on larger rated SCR Power Controllers, fans aid the heatsink for heat removal. Present day fans are very reliable and do not present any particular disadvantage to the user. Check the blades for dirt and grime build-up and check the fan for proper operation.

If the circuit boards have edge connectors or multi-pin connectors, check each of them for cleanliness, tightness and proper fit. If these do not seem proper, contact the manufacturer.

If the SCR Power Controller is water cooled, then the tightness of the water connections should be checked periodically. In addition, the water quality should be checked in accordance with the information in Chapter 9.

TORQUE REQUIREMENTS

Refer to the chart in Figure 33 for the proper torque.

Screw/Bolt Size	Torque Value
6-32	Tight to the feel
8-32	15 inch-pounds
10-32	30 inch-pounds
1/4-20	70 inch-pounds
5/16-18	100 inch-pounds
3/8-16	375 inch-pounds
1/2-13	500 inch-pounds

Figure 33 - Torque Ratings Chart

If a mechanical lug such as those manufactured by Burndy Corporation or ILSCO is used, check the manufacturer's data sheet for the tightness information.

Figure 34 - Mechanical Lug Examples
(photo courtesy of ILSCO)

SUMMARY

The following items are essential to overall life of the SCR Power Controller.

* Check and tighten all electrical connections
* Clean corrosion from connections if needed
* Clean the heatsink
* Check and clean/replace enclosure air filters
* Clean and check operation of heatsink fans
* Check edge connectors and multi-pin connectors
* If water cooled, check and tighten water connections

If each of these maintenance items is performed periodically, you should have years of trouble free operation from the SCR Power Controller.

Chapter 12

THE COMPLETE INDUSTRIAL
ELECTRIC HEATING SYSTEM

GENERAL

The modern day industrial electric heating system takes on many
different shapes, sizes, configurations and has many different
applications. Of course, there are dozens of manufactures which
design and build specialty industrial electric heating systems.
Refer to Chapter 2 for a partial listing of applications. One
thing remains the same for all of these different systems - they
all have the same main components.

In my opinion, electric heat is the best overall solution for
industrial heating applications. The use of electricity is

clean, safe, environmentally friendly, has excellent controllability and the cost is more stable since the price of fossil fuels varies so much from day to day in the world markets.

MAIN COMPONENTS

The main components include the Furnace Shell, Heating Elements, SCR Power Controls, Temperature Controls, Temperature Sensors and Motion Control.

Furnace Shell

The Furnace Shell is the containment area in which the heating takes place. The size and shape is determined by many factors including the following:

- Type, Size, Weight of Product to be heated
- Quantity of product to be heated
- Material product is made of
- Reason for heating such as drying, hardening, etc.

The Furnace is designed and built once all of this information is available to the engineer.

Heating Elements

The Heating Element is selected mainly based upon Furnace temperature requirements and the available electrical power. Heating Elements are made from standard materials such as Nichrome wire to exotic materials such as Molybdenum-Disilicide. The shape and size of Heating Elements are available in standard formats or can be custom designed to fit the specific application.

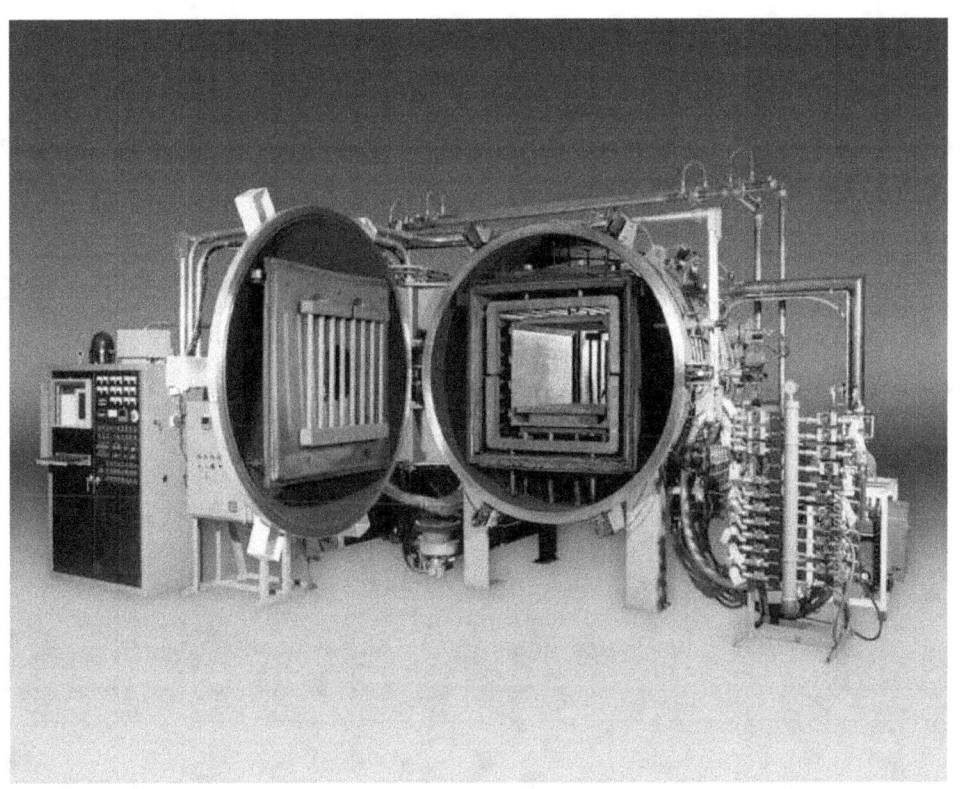

Figure 35 – Centorr Vacuum Industries Sintervac ® C25, Vacuum and Controlled Atmosphere Debind and Sintering Furnace. Rated for 2300C with Graphite Heating Elements and Hot Zone
(photo courtesy of Centorr Vacuum Industries)

SCR Power Control

The SCR Power Control actually controls the power to the heating elements. The SCR responds to a command signal typically 4-20ma from a Temperature Controller. If the Furnace temperature is above the desired level, the SCR reduces its output. If the Furnace temperature is below the desired level, the SCR increases its output. As described earlier in this book, the SCR Power Controller is available in many shapes, sizes, control methods and voltage/current ratings. The exact SCR is specified based upon the input voltage, number of phases, kilowatt requirements and heating element type.

Temperature Controllers and Sensors

The Temperature Control monitors the Furnace's temperature via a Thermocouple, Infra-red sensor or one of many other types of temperature sensors. It converts this signal into a dc signal that is fed to the SCR. This signal is inversely proportional to the actual temperature. In other words, as the temperature increases, the Temperature Controller's output decreases. As the temperature decreases the output increases. The SCR responds to this signal.

Temperature Controllers have Proportional, Integral and Derivative (PID) control algorithms. Not many years ago the Furnace user had to adjust these control loops to obtain optimum Furnace performance. However, modern Temperature Controls have "Auto PID" tuners. These tuners learn the process very quickly and make the appropriate adjustments in the Temperature Controller's memory. This feature has greatly reduced the initial start-up time of a new Furnace.

Proportional is the gain setting of the Temperature Controller. The lower the gain (proportional) the further the process temperature will be from the desired temperature. Increasing the gain brings the actual and the desired temperatures closer together (commonly called the offset). However, the higher the gain, the more likely you are to have oscillations.

The Integral loop (sometimes called reset) slows the oscillations and works to narrow the offset in temperatures.

Finally, the Derivative loop (frequently called the rate). This is the most difficult loop to adjust and many engineers simply don't use it because it can cause serious stability problems for the Furnace's temperature. Most engineers stick to the 2 mode or PI controller since it is so much easier to set up.

As mentioned earlier, the "auto tune" controllers minimize

the overall knowledge of a particular furnace operation and ease the set-up of the furnace.

Motion Control

Many Furnaces require Motion Control to move the products in and out of the Furnace. This can be anything such as a conveyor or complicated winders such as used for wire and cable annealing. The speed of the conveyor or winder can affect the required heating requirements.

TYPICAL HEATING SYSTEM

The Typical Heating System can have one or more heating zones. For example, Large Furnaces may require more than one heating zone. This could be the result of a Furnace being very long. Another example is a Furnace in which the first heating zone pre-heats the product. The next heating zone maintains a constant temperature for a predetermined length of time. The final heating zone could be used for cooling by providing a smaller amount of heat to the product to actually slow the cooling process. Of course there are many variations on this scheme.

SUMMARY

Industrial Electric Heating Systems are used in many industrial applications. Electric heat is more user friendly and environmentally safe as well. These systems can be small, medium or large in size. They can also be relatively simple or very complex depending on the application. The use of "auto tune" Temperature Controllers can ease start-up of a new furnace.

Chapter 13

WHAT THE FUTURE HOLDS FOR SCR POWER CONTROLLERS

GENERAL COMMENTS

I've never considered myself to be a visionary or prophet and don't claim to be one! However, after working in this industry for over 30 years I've experienced many changes and I believe I have a good idea of what this product has in its future.

Many engineers have said for years the SCR is something of the past. Colleges do little to educate students about SCRs and/or their uses.

With all of this said and over fifty years since the SCRs inception, it is still the workhorse of the power control industry.

RATINGS

SCR ratings will continue to increase. Better techniques for growing silicon will yield higher current ratings. The use of better silicon slicing machines, encapsulation techniques and clean rooms will improve manufacturing and yield higher voltage ratings. New housing materials may help yield higher ratings also.

Gate designs will continue to change and mature. These improvements will make "turning on" larger diameter SCRs easier and easier. "Built in" optical gating will make large SCR controlled Power Supplies easier to design and build.

PACKAGING

The manufactures will continue to find new ways and techniques for providing a "packaged" array of SCRs to help improve mechanical design problems. Overall Size has always been a concern. New SCR packages will continue to help shrink the SCR Power Controller.

ELECTRONIC CIRCUITS

Larger scale Integrated Circuits continue to increase in complexity and the cost of custom ICs have become lower. As this trend continues, SCR Power Controllers will utilize custom ICs for control. Again this will help make the complete unit smaller, more reliable and less expensive to build. Figure 36 illustrates a three-phase SCR Firing Circuit designed and built by Enerpro, Inc. Enerpro was the first SCR Power Control manufacturer to use custom manufactured integrated circuits.

**Figure 36 – Three-Phase SCR Firing Circuit
with custom integrated circuits**
(photo courtesy of Enerpro, Inc.)

The use of these custom IC's has allowed Enerpro to manufacture
a circuit that can be used in a wide range of SCR power circuit
topologies.

COMMUNICATIONS

In 1989 I wrote specifications and began work on an SCR Power
Controller with digital communications. I really felt that was
the wave of the future. I left that company in 1991 ad the
project was scraped shortly thereafter.

Here it is over 15 years later and Power Control manufacturers
are just starting to add digital communications. The biggest

problem these manufactures are faced with is what communication protocol to use. They haven't figured it out yet and I'm not sure there is a correct answer.

The use of digital communications allow the user to adjust output levels, select firing modes, adjust feedback and current limit levels, monitor alarms and input/output voltage, current and power levels.

It seems to me that manufacturing processes utilizing many individual zones of heat control are the likely users of SCR Power Controllers with digital communications. The initial cost of the Power Control equipment may be higher however, the reduced installation cost will more than offset for this additional cost. In addition, data logging of the setpoints, output parameters and alarms may help the user to better understand his or her furnace operation. This in turn could lead to improved processes, cost reductions due to minimizing waste and ultimately the user could make a better product.

SUMMARY

In my mind, the SCR Power Controller is here to stay. Many improvements will be made to the discrete device and to the SCR Power Controller. Sizes will shrink and power ratings will increase. Additional features may make the SCR Power Controller the right power control device for applications not even thought of today.

Refer to the "Glossary of Terms" section for general terminology clarifications immediately following this chapter. For more information on the use and applications of SCR Power Controllers please refer to the "Bibliography" section located at the end of this book.

GLOSSARY OF TERMS

AC SWITCH - consists of two SCRs connected in an inverse parallel or back to back configuration. This is used to control AC current.

AMBIENT TEMPERATURE - this is the temperature in which an SCR Power Control is expected to operate in. This is usually 0° to 50°.

ANODE - the negative power terminal of an SCR or Diode.

CATHODE - the positive power terminal of an SCR or Diode.

COMMAND SIGNAL - a variable input to an SCR Power Control that determines the output setting. It can be a current, voltage, potentiometer or slide-wire input.

CSA - This stands for Canadian Standards Association. This organization provides 3rd party certifications of a product to Canadian Standards.

cUL - Underwriters Laboratories' mark for 3rd party certification of a product to Canadian Standards.

CURRENT LIMITING - the means for limiting or setting the maximum current level applied to the load. This should be RMS for best results.

CURRENT REGULATION - the means for regulating the current to a ever-changing load resistance. It also linearizes the input command signal to the output current. This should be RMS for best results.

CURRENT TRANSFORMER - commonly referred to as a **CT**. It is used to sense the AC current. Its output is isolated and linear to the measured current. The output is normally 0 to 5 amps.

DC COMPONENT - a small amount of DC current (typically milli-amps) present in the output of a SCR Power Controller. It is caused by the unequal timing or phase angle of the positive and negative half cycle of the AC waveform. It can cause problems if it is large.

di/dt - the rate of rise of applied current to an SCR as it turns on or starts to conduct.

DIODE - a semiconductor that allows current to flow in only one direction. It has no control terminal.

dv/dt - the maximum rate of rise of applied voltage across an "off" SCR that will not cause a false turn-on.

ENCLOSURE - the "box" in which an SCR Power Control will be installed - Usually a NEMA type such as a NEMA 1.

FORWARD DROP - this is the voltage drop across an SCR or Diode during conduction in the normal forward direction. This voltage drop multiplied times the current determines the watt loss of the SCR or Diode. It is typically 1.3 to 1.5 volts.

GATE - the control terminal of an SCR used to turn it on.

HARMONICS - are generated when electrical energy is switched and is normally undesirable. Harmonics are expressed as a multiple of the fundamental frequency.

HEATING ELEMENT - this is the electrical device that produces heat when electrical current is passed through it. It is selected by the temperature requirements, voltage & current ratings and the mechanical shape.

HEATSINK - a mechanical device used to transfer heat from the SCR or Diode. It can be convection, fan or water cooled.

HYBRID CIRCUIT - refers to a SCR and Diode combination in an SCR Power Control; normally a 3 phase unit, 3 SCRs and 3 Diodes. It has limitations in certain applications such as unbalance loads.

A 6 SCR unit will work in any application that a hybrid was previously used.

INRUSH CURRENT - the current that surges when a low resistance load is energized or the current drawn by a transformer during initialization or saturation.

LED - this stands for Light Emitting Diode. Typically used as diagnostic indicators on electronic equipment.

METAL OXIDE VARISTOR (MOV) - A device used in conjunction with an R-C Snubber circuit to protect the SCR or Diode from voltage transients.

OVERCURRENT TRIP - this is a electronic circuit which monitors and "shuts down" an SCR if the peak current has exceeded a preset level. It is sometimes referred to as "Electronic Fusing".

PEAK - the maximum instantaneous value of voltage or current.

PHASE-FIRING - one of two firing modes for SCRs. Each SCR is turned on for only a portion of each half of an AC cycle. It can cause a poor power factor.

PHASE-LOCK-LOOP - an electronic circuit that automatically adjusts itself to maintain synchronization with the line frequency. It can also be used as a line frequency noise filter.

PIV - This stands for Peak Inverse Voltage and is the voltage rating of an SCR or Diode. It should be approximately 2.3 times the operating AC RMS voltage.

POTENTIAL TRANSFORMER - usually referred to as a PT. Used to isolate or change a voltage level typically for metering purposes.

POWER FACTOR (P.F.) - the utilization factor of the power being consumed from the power source. A poor power factor means the user is paying for energy not being used to produce heat for example.

R-C SNUBBER - series connected resistor and capacitor network connected across an SCR to "slow down" the rate of applied voltage (dv/dt) to help prevent the SCR from falsely turning on. This is used in conjunction with a MOV for maximum protection.

RADIO FREQUENCY INTERFERENCE (RFI) - this is high frequency electrical interference generated by the chopping action of phase-fired SCRs. Very little RFI is typically produced by zero-fired SCRs.

RMS - this stands for Root Mean Squared, refers to the heating value of current or voltage.

SCR POWER CONTROL - an electronic control device that uses discrete SCRs as the control device and are typically used in industrial electrical heating applications.

SCR - Silicon Controlled Rectifier, a device used to switch the power to the load in an SCR Power Control.

SEMICONDUCTOR FUSE - the protective device used to save an SCR in the case of an overcurrent condition. It is usually a very fast acting fuse (see I^2T). Not for wire/cable protection.

SOFT-START - a ramping effect of voltage or current to the load to minimize or eliminate inrush currents.

SOLIDSTATE RELAY (SSR) - a module which contains two SCRs with its own isolation circuitry built in.

SPAN ADJUSTMENT - a multi-turn potentiometer for matching an SCR Power Control's output to its command signal. It is usually adjusted with the command signal at 100%. It is sometimes referred to as the "Gain" adjustment.

THERMAL IMPEDANCE - the value placed on a heatsink indicating it's capability to remove heat. The lower the number the better the heat removal is. It is usually expressed as °C/Watt.

THYRISTOR - see SCR.

UL - Underwriters Laboratories. This organization provides 3rd party certification of products. The mark indicates the product meets the stringent safety specifications. It is commonly required for installation electrical equipment in large cities such as Chicago or New York City.

VARIABLE-TIME-BASE - a control mode for zero-fired SCR Power Controls. The ratio of the number of on and off AC cycles is constantly changing to minimize power fluctuations to the load. This helps minimize thermal shock to the heating elements, which helps increase the element's life expectancy.

VOLTAGE REGULATION - a means for regulating the output voltage of an SCR Power Control or for compensating for a changing input line voltage. This should be RMS for best results.

ZERO ADJUSTMENT - a multi-turn potentiometer for matching an SCR Power Control's output to its command signal. This is usually adjusted for zero output when the command signal is at zero. It is sometimes referred to as the "Bias" adjustment.

ZERO-FIRED - a method of controlling power to the load with full sinewave outputs, maintains a near unity power factor. The SCR is always turned on at its zero crossing. It minimizes harmonics.

BIBLIOGRAPHY

"Power Converter Handbook"
(Canadian General Electric Company Limited)

"What You Should Know About SCR Power Controllers"
(Control Concepts)

"Power Electronics and AC Drives"
B.K. Bose.
(Prentice-Hall, Englewood Cliffs, New Jersey 07632)

"Selecting Power Controls"
R. Brad Vogelbach
(Process Heating Magazine - June 1997)

"SCR Manual - 5th edition"
(General Electric Co., Auburn, N.Y., 1972)

"GLOBAR Silicon Carbide Electric Heating Elements"
(The Carborundum Company, Niagara Falls, N.Y. 14302)

"Kanthal Handbook"
(The Kanthal Corporation, Bethel, CT 06801, 1989)

"IEEE Std 519-1992 - IEEE Recommended Practices and Requirements for Harmonic Control in Electrical Power Systems"
(Institute of Electrical and Electronics Engineers, Inc, New York, N.Y. 10017)

"Selecting an Enclosure for SCR Power Controls"
George A. Sites
(Industrial Heating Magazine - May 1991)

"Application Note - 1004 Water as an Alternative Method for Cooling High Power Electronic Equipment"
George A. Sites
(HDR Power Systems, Inc., Columbus, Oh. 43204)

"Application Note - 1005 Commonly Used Terms with SCR Power Controls"
George A. Sites
(HDR Power Systems, Inc., Columbus, Oh. 43204)

"Controlling Silicon Carbide Heating Elements with SCR Power Controls"
George A. Sites
(Industrial Heating Magazine - September 1997)

"Electric Fuses"
A. Wright & P.G. Newbery
(Peter Peregrinus LTD., London, U.K. 1982)

"Controller Tuning & Control Loop Performance - A Primer"
David W. St. Clair
(Straight-Line Control Company, Newark, Delaware 1989)

INDEX

A

advantages, 2, 3, 33, 34
ambient temperature, 69, 70, 71, 72
Anode, 22, 27

C

Cathode, 22, 27
Centorr Vacuum Industries, 9, 82, 83
command signal, 15, 40, 42, 43, 83, I, IV
communications, 89
Communications, xvii
COMMUNICATIONS, 89
Crouzet, 29, 30

D

DC component, 16
Diode, 22, 37, I, II, III
disadvantages, 2, 3, 33, 37
DISADVANTAGES, 3
Diversified Controls & Systems, Inc., 10
duty cycle, 43, 63, 72
duty-cycle, 64

E

electro/mechanical, 77
Enerpro, Inc., 89

F

Fan, xxi, 64, 75
fans, 30, 64, 78, 79
feedback, 42, 43, 45, 56, 57, 89
FEEDBACK, xvi, 42

RMS CURRENT REGULATION, 44
RMS Voltage Regulation, xvi
RMS VOLTAGE REGULATION, 44

S

saturable core reactor, 2
SCR, i, iii, v, vii, xv, xvii, xx, 1, 2, 3, 6, 7, 8, 11, 13, 14, 15, 16, 17, 18, 19, 21, 22, 23, 24,
    25, 26, 27, 28, 29, 30, 31, 32, 33, 34, 36, 37, 42, 43, 44, 45, 47, 48, 49, 50, 51, 52, 53, 55,
    57, 58, 60, 62, 63, 64, 68, 70, 75, 77, 78, 79, 82, 83, 87, 88, 89, 90, I, II, III, IV, V, VI, VII
SCR Power Control, vii, xx, 2, 3, 6, 11, 13, 22, 26, 34, 36, 45, 49, 50, 55, 57, 63, 64, 68, 70,
    75, 83, 89, I, II, IV
SCR Power Controls, 1, 2, 3, 7, 8, 14, 22, 28, 33, 37, 42, 43, 45, 47, 48, 49, 53, 62, 82, IV,
    VI, VII
SCR POWER CONTROLS, *i, iii, v, xv*
semiconductor fuse, 17, 25
Semiconductor Fuse, 17
Silicon Carbide, xxi, 56, 57, VI, VII
Silicon Controlled Rectifier, 21, IV
SILICON CONTROLLED RECTIFIER, 13
Silicon Controlled Rectifiers, 13
snubber, 17, 25
Snubber, xx, 17, 18, 24, 29, III
Soft-Start, 39, 40
solid-state, 2, 14, 21, 29, 32
Solidstate Relay, 26
Solidstate Relays, xx
SSR, xx, 26, 28, 29, IV
Stud Mount, xx, 26, 27

T

Temperature Control, 83
Temperature Controllers, 83, 85
Thyristor, 21
TORQUE, 78
Transient, xx, 17, 18
True Power Regulation, xvi, 45

www.ingramcontent.com/pod-product-compliance
Lightning Source LLC
Chambersburg PA
CBHW081456170526
45166CB00008B/2446